Methods for Real-Time Measurement of THMs and HAAs in Distribution Systems

The mission of the Awwa Research Foundation (AwwaRF) is to advance the science of water to improve the quality of life. Funded primarily through annual subscription payments from over 1,000 utilities, consulting firms, and manufacturers in North America and abroad, AwwaRF sponsors research on all aspects of drinking water, including supply and resources, treatment, monitoring and analysis, distribution, management, and health effects.

From its headquarters in Denver, Colorado, the AwwaRF staff directs and supports the efforts of over 800 volunteers, who are the heart of the research program. These volunteers, serving on various boards and committees, use their expertise to select and monitor research studies to benefit the entire drinking water community.

Research findings are disseminated through a number of technology transfer activities, including research reports, conferences, videotape summaries, and periodicals.

Methods for Real-Time Measurement of THMs and HAAs in Distribution Systems

Prepared by:
Gary L. Emmert, **Gang Cao**, **Gija Geme**, **Neena Joshi**, and **Mostafizur Rahman**
The University of Memphis
Department of Chemistry
Smith Chemistry Building
Memphis, TN 38152

Jointly sponsored by:
Awwa Research Foundation
6666 West Quincy Avenue
Denver, CO 80235-3098

and

U.S. Environmental Protection Agency
Washington, D.C.

Published by:

Awwa Research Foundation
Advancing the Science of Water®

IWA Publishing

DISCLAIMER

This study was funded by the Awwa Research Foundation (AwwaRF). AwwaRF assumes no responsibility for the content of the research study reported in this publication or for the opinions or statements of fact expressed in the report. The mention of trade names for commercial products does not represent or imply the approval or endorsement of AwwaRF. This report is presented solely for informational purposes.

Published by IWA Publishing, Alliance House, 12 Caxton Street, London SW1H 0QS, UK

Telephone: +44 (0) 20 7654 5500; Fax: +44 (0) 20 7654 5555; Email: publications@iwap.co.uk;
Web: **www.iwapublishing.com**

AwwaRF report number 91003F. Originally published by AwwaRF for its subscribers in 2003.
IWA Publishing version published 2004.

ISBN 1 84339 8982

CONTENTS

LIST OF TABLES vii

LIST OF FIGURES ix

FOREWORD xi

ACKNOWLEDGMENTS xiii

EXECUTIVE SUMMARY xv

CHAPTER 1: TRIHALOMETHANES AND HALOACETIC ACIDS IN DRINKING WATER 1
- Overview 1
- Research Objectives 2
- Significance of Project 2
- Expected Outcomes 2

CHAPTER 2: OVERVIEW OF EXISTING ANALYTICAL METHODS FOR MONITORING THMs IN DRINKING WATER DISTRIBUTION SYSTEMS 5
- Environmental Protection Agency Methods 5
- Alternative THMs Methods Appearing in the Literature 5
  - GC Based Methods for THMs 5
  - GC-MS Based Methods 22
  - Colorimetric Methods 29
- Summary of THM Methods 33

CHAPTER 3: OVERVIEW OF EXISTING ANALYTICAL METHODS FOR MONITORING HAAs IN DRINKING WATER DISTRIBUTION SYSTEMS 35
- Environmental Protection Agency Methods 35
- Alternative Methods for HAAs 35
  - GC Based Methods for HAAs 35
  - GC-MS Based Methods for HAAs 40
  - SPME Based Methods for HAAs 44
  - HPLC / IC Based Methods for HAAs 46
  - LC-MS Methods for HAAs 50
  - Capillary Electrophoresis Methods for HAAs 52
  - Colorimetric Based Methods for HAAs 56
  - Electrospray Ionization-High-Field Asymmetric Waveform Ion Mobility Spectrometry-MS 57
  - Other Notable Sample Preparation/Introduction Methods 58
- Summary of HAA Methods 59

CHAPTER 4: RECOMMENDATIONS FOR FURTHER RESEARCH 61
Conclusions for the Final Report 61
Defining the Ideal Method 62
Summary of Methods for THMs Analysis 63
Summary of Methods for HAAs Analysis 65

REFERENCES 93

ABBREVIATIONS 101

TABLES

2.1 Reported MDL, accuracy and precision results for THMs using USEPA Methods 33

2.2 Types of commercially available SPME fibers 34

2.3 Comparison of the MDL, accuracy and precision studies for the SCMS-GC-ECD and the SCMS-GC-PDECD methods for measuring THM concentrations in water 34

3.1 Expected MDL, accuracy and precision results for HAAs using USEPA Methods 59

4.1 Summary of analytical methods from the literature for measuring THMs concentrations in drinking water 67

4.2 Summary of analytical methods from the literature for measuring the concentrations of five HAAs in drinking water 76

4.3 The GC columns used for separation of THMs 88

4.4 The GC columns used for separation of HAAs 89

4.5 Internal standards and surrogate analytes 90

4.6 The LC columns used for separation of HAAs 91

FIGURES

2.1 Schematic diagram of the laboratory built apparatus for P&T 8

2.2 Diagram of the dynamic HS sampling device 10

2.3 Diagram of an SPME sampling assembly. 13

2.4 The SCMS sampling probe 17

2.5 The SCMS probe and calibration vessel in the calibration configuration 18

2.6 The SCMS-GC analyzer 18

2.7 Fast-loop/flow-cell configuration 19

2.8 The classical pathway for the Fujiwara reaction 30

2.9 Schematic diagram of continuous flow-FIA analyzer with fluorometric detection 31

FOREWORD

The Awwa Research Foundation is a nonprofit corporation that is dedicated to the implementation of a research effort to help utilities respond to regulatory requirements and traditional high-priority concerns of the industry. The research agenda is developed through a process of consultation with subscribers and drinking water professionals. Under the umbrella of a Strategic Research Plan, the Research Advisory Council prioritizes the suggested projects based upon current and future needs, applicability, and past work; the recommendations are forwarded to the Board of Trustees for final selection. The foundation also sponsors research projects through an unsolicited proposal process; the Collaborative Research, Research Applications, and Tailored Collaboration programs; and various joint research efforts with organizations such as the U.S. Environmental Protection Agency, the U.S. Bureau of Reclamation, and the Association of California Water Agencies.

This publication is a result of one of these sponsored studies, and it is hoped that its findings will be applied in communities throughout the world. The following report serves not only as a means of communicating the results of the water industry's centralized research program but also as a tool to enlist the further support of the nonmember utilities and individuals.

Projects are managed closely from their inception to the final report by the foundation's staff and large cadre of volunteers who willingly contribute their time and expertise. The foundation serves a planning and management function and awards contracts to other institutions such as water utilities, universities, and engineering firms. The funding for this research effort comes primarily from the Subscription Program, through which water utilities subscribe to the research program and make an annual payment proportionate to the volume of water they deliver and consultants and manufacturers subscribe based on their annual billings. The program offers a cost-effective and fair method for funding research in the public interest.

A broad spectrum of water supply issues is addressed by the foundation's research agenda: resources, treatment and operations, distribution and storage, water quality and analysis, toxicology, economics, and management. The ultimate purpose of the coordinated effort is to assist water suppliers to provide the highest possible quality of water economically and reliably. The true benefits are realized when the results are implemented at the utility level. The foundation's trustees are pleased to offer this publication as a contribution toward that end.

Edmund G. Archuleta, P.E.
Chair, Board of Trustees
Awwa Research Foundation

James F. Manwaring, P.E.
Executive Director
Awwa Research Foundation

FOREWORD

ACKNOWLEDGMENTS

The authors of this report are indebted to the following water utilities and organizations for their cooperation and participation in this project.

- Memphis Light Gas and Water (MLGW), Water Quality Assurance Laboratory, Memphis, Tennessee
- City of Houston, Public Works and Engineering Department, Water Production Branch, Houston, Texas
- The Department of Chemistry, College of Arts and Sciences and Office of the Vice-Provost, The University of Memphis, Memphis, Tennessee

The authors would like to extend thanks to Alice Fulmer of Awwa Research Foundation who served as Project Manager for the majority of this project. Alice has provided excellent guidance for the authors and the Project Advisory Committee (PAC). We appreciate her hard work in keeping the project on track. The authors also extend thanks to Kenan Ozekin who served as the initial Project Manager for this study. The authors are also indebted to the members of the PAC who provided exceptional advice during the review of this Report. The PAC members include: Bill B. Potter, Research Chemist at the USEPA, Cincinnati, OH; Claudia Wheeler, Metropolitan Water District of Salt Lake City, Sandy, UT; and Yuefang Xie, Associate Professor, Penn State University-Harrisburg, Middletown, PA. Appreciation is also is extended to Torrence Myers, Manager of Water Quality Assurance Laboratory of MLGW and Ying Wei, Assistant Laboratory Chief, Water Production Branch, City of Houston, Public Works and Engineering Department.

In addition, Michael Brown and Anshon Jenkins of the Department of Chemistry, The University of Memphis provided invaluable technical assistance with the preparation and editing this report.

EXECUTIVE SUMMARY

Chlorination is the most widely used water disinfection process in the United States. Chlorine, typically applied as gaseous chlorine, sodium hypochlorite, or calcium hypochlorite is a safe and economical water disinfectant. Chlorination has been used with great success in the water industry for over a century. Unfortunately, water chlorination results in the formation of halogenated disinfection by-products (DBPs). These DBPs include the trihalomethanes (THMs) and haloacetic acids (HAAs).

The possibility of adverse health effects of THMs and HAAs has led to the establishment of maximum contaminant levels (MCLs) in drinking water. The THMs include chloroform, bromodichloromethane, dibromochloromethane and bromoform. The current MCL for total THMs in drinking water is 0.080 mg/L. The five HAAs that are commonly monitored include monochloroacetic acid, dichloroacetic acid, trichloroacetic acid, monobromoacetic acid and dibromoacetic acid and are referred to as HAA5. The current MCL addresses HAA5 and is set at 0.060 mg/L for total concentration of HAA5.

Stage Two of the Disinfectant and Disinfection By-Product Rule may lower the MCL for total THMs and HAAs in the near future. This will require many plant operators in water utilities to develop new strategies for minimizing total THMs and total HAAs concentrations in drinking water. They might also look toward optimizing their removal processes. To accomplish these goals, analytical methods are needed that will provide monitoring data in real-time or near-real time. The development of such measurement methods for on-line monitoring will improve understanding of variations in THM and HAA concentrations in treatment plants and drinking water distribution systems. The monitoring data will increase understanding of how peak THM and HAA formation events occur (and when they do not occur). This allows operators to adjust treatment processes accordingly at the plant and in the distribution system. Finally, the data that such monitoring provides can be used to develop better models of the water distribution system.

RESEARCH OBJECTIVES

The objectives of Phase One are to conduct an extensive survey of the literature to evaluate which existing analytical methods may be adapted for real time monitoring of THMs and HAAs in drinking water distribution systems. These methods will be compared to the U.S Environmental Protection Agency (USEPA) Methods for monitoring these disinfection by-products in drinking water. The methods that will be examined/adapted may include USEPA Methods 502.2, 524.2, 551 and 551.1 for THMs and USEPA Methods 552, 552.1, 552.2 and 552.3 for HAAs. All methods will be evaluated as to their potential to be applied and /or adapted for on-line monitoring of the concentrations of THMs and HAAs expected in water distribution systems.

APPROACH

While THMs and HAAs have in common that they are disinfection by-products of water chlorination, they are very different in their chemical structures and properties. The THMs are non-polar and volatile. The HAAs are polar and non-volatile. These differences have significant implications with regard to their analytical chemistry applied to drinking water, particularly the sampling-sample preparation step. The concentrations of THMs can be determined using gas chromatography. The major advantage of gas chromatography (GC) is its simplicity. GC methods are relatively easy to use and automate and standard versions are well established. GC methods are relatively inexpensive and produce minimal waste. The analysis of the HAAs using GC is more complicated. To be analyzed by GC, HAAs must be converted to their corresponding methyl esters before they are volatile enough for routine GC analysis. Typically, this adds manual steps to the sample preparation process. Many of these steps are susceptible to transfer losses or carry-over of the analyte.

Much effort has been exerted to minimize these sample preparation steps required by GC. Liquid chromatographic (LC) techniques such as high performance liquid chromatography (HPLC), ion chromatography (IC), and liquid chromatographic-mass spectrometer (LC-MS) methods have been developed for HAAs. While to many, LC based analysis may seem more complex, the reality is that these methods are becoming well established and routine. The cost of the technology has dropped considerably. Some of the LC based methods for HAAs may be applied with minimal investment.

CONCLUSIONS FOR FINAL REPORT FOR PHASE ONE

The literature provides much information pertaining to the analysis of THMs and HAAs in drinking water. Over 100 papers have been reviewed and summarized in this Report. The goal was to collect data about the specific operating parameters used in the methods so that they could be evaluated as to their adaptability for on-line real-time monitoring in drinking water distribution systems. Several of the reviewed methods show promise and include a wide array of methodologies ranging from simple colorimetric spot tests to more complex LC-MS methods and beyond. Careful review of the literature continues in our laboratory.

An electronic database has been submitted in rich text format (RTF) that is searchable via the "find" feature of most commercially available word processors. This database includes references to all THMs and HAAs papers reviewed to date. It is the hope of the authors that this database will be a valuable resource for the AwwaRF members and researchers interested in analytical method development.

The concept of the Ideal Method was defined. A discussion of the Ideal Method included a description of the desirable characteristics of the Ideal Method or Methods for THM and HAA analysis as applied to real-time monitoring in distribution systems.

Two tables designed to summarize the results of the comprehensive literature review are included in Chapter 4. Table 4.1 summarizes the THM methods. Table 4.2 summarizes HAA methods. The tables provide details of each method including: method detection limit, accuracy and precision data; the commercial availability of instrumentation; EPA basis of the method; estimated instrumentation and skill level of the analyst; estimated instrumentation cost; estimate of instrument ruggedness and portability; the estimated analysis time and sample analysis rate; the ability to adapt to on-line monitoring as a real-time or near real-time measurement; and possible automation modes for the method. Additional tables provide a resource to readers and collect pertinent information related to the methods. Tables 4.3 and 4.4 list the common commercially available GC columns that were used for THM and HAA analysis, respectively. Table 4.5 lists the common internal and surrogate standards that were used for THM and HAA methods. Table 4.6 lists the common HPLC / IC columns that are used for HAA analysis. Several options are presented describing the possibilities for automating THM and HAAs analysis for real-time monitoring in distribution systems.

CHAPTER 1
TRIHALOMETHANES AND HALOACETIC ACIDS IN DRINKING WATER

OVERVIEW

Chlorination is the most widely used water disinfection process in the United States. Chlorine, typically applied as gaseous chlorine (Cl_2), sodium hypochlorite (NaOCl), or calcium hypochlorite ($Ca(OCl)_2$), is a safe and economical water disinfectant. Chlorination has been used with great success in the water industry for over a century. Unfortunately, water chlorination results in the formation of halogenated disinfection by-products (DBPs). These DBPs include trihalomethanes (THMs) and haloacetic acids (HAAs).

There is concern over the possible health effects of THMs and HAAs . This concern has led to maximum contaminant levels (MCLs) being established in drinking water. The THMs include chloroform ($CHCl_3$), bromodichloromethane ($CHBrCl_2$), dibromochloromethane ($CHBr_2Cl$) and bromoform ($CHBr_3$). The current MCL for total THMs in drinking water is 0.080 mg/L.

The five HAAs that are commonly monitored include monochloroacetic acid (MCAA), dichloroacetic acid (DCAA), trichloroacetic acid (TCAA), monobromoacetic acid (MBAA) and dibromoacetic acid (DBAA) and are referred to as HAA5. There are other possible chlorinated and brominated HAAs of interest in drinking water. For example, HAA6 includes the species in HAA5 and bromochloroacetic acid (BCAA). HAA9 includes HAA6, bromodichloroacetic acid (BDCAA), chlorodibromoacetic acid (CDBAA) and tribromoacetic acid (TBAA). The current MCL addresses HAA5 and is set at 0.060 mg/L for total concentration of HAA5.

It was thought that Stage Two of the Disinfectant and Disinfection By-Product (D/DPB) Rule (USEPA 1998) might lower the MCL for total THMs and HAAs in the near future. However, more recent proposals indicate that these MCLs will remain in effect (USEPA 2003c). Regardless, many plant operators in water utilities will want to develop new strategies for minimizing THM and HAA concentrations in drinking water. They might also look toward optimizing THM and HAA removal processes.

To accomplish these goals, analytical methods are needed that will provide data on THM and HAA concentrations in real-time or near-real time. The development of such measurement methods for on-line monitoring will improve understanding of variations in THM and HAA concentrations in treatment plants and drinking water distribution systems. The monitoring data will increase understanding of how peak THM and HAA formation events occur (and when they do not occur), allowing operators to adjust treatment processes accordingly at the plant and in the distribution system. Finally, the data that such monitoring provides can be used to develop better models of the water distribution system.

While THMs and HAAs have in common that they are both DBPs of chlorine, they are very different in their chemical structure and properties. The THMs are non-polar and volatile while the HAAs are polar and non-volatile. These differences have significant implications in their analytical chemistry applied to drinking water, particularly the sampling-sample preparation step. The concentrations of THMs can be determined using gas chromatography (GC). The major advantage of GC is its simplicity. GC is easy to use and automate, and these methods are well established. GC is also inexpensive and produces minimal waste. The analysis of the HAAs using GC is more complicated. To be analyzed by GC, HAAs must be converted to their corresponding methyl esters before they are volatile enough for routine GC analysis. Typically this adds manual steps to the sample preparation process. Many of these steps are susceptible to transfer losses or carry-over of the analyte.

Much effort has been exerted to minimize these sample preparation steps required by GC. Liquid chromatographic (LC) techniques such as high performance liquid chromatography (HPLC), ion chromatography (IC) and HPLC-mass spectrometry (MS) methods have been developed for HAAs. While to many, HPLC and IC analysis may seem more complex (meaning it requires a higher level of analytical expertise to obtain reliable results), the reality is that these methods are becoming well established and routine. The cost of the technology, i.e., high-pressure pumps, injection valves and simple detectors, has dropped considerably. Some of the LC based methods for HAAs can be applied with minimal investment. Because of the differences in the chemical properties and the analytical approach, the methods for measuring THMs and HAAs will be considered separately.

There are several very helpful review articles (Urbansky 2000a, Urbansky 2000b, Weinberg et al. 1999, Richardson 2001a, Richardson 2001b, Richardson 2002, Reemtsma 2001) appearing in the literature that focus on the application of more advanced instrumental techniques for DBPs monitoring. This Report will include a broader prospective that includes many of these advanced instrumental methods as well as simpler and generally more inexpensive methods.

RESEARCH OBJECTIVES

The objectives of Phase One are to conduct an extensive survey of the literature to evaluate which existing analytical methods may be adapted for real time monitoring of THMs and HAAs in drinking water distribution systems. These methods will be compared to the USEPA Methods for monitoring THMs and HAAs in drinking water. The methods that may be examined/adapted include USEPA Methods 502.2 (USEPA 1995a), 524.2 (USEPA 1995b), 551 (USEPA 1990a) and 551.1 (USEPA 1995c) for THMs and USEPA Methods 552 (USEPA 1990b), 552.1 (USEPA 1992), 552.2 (USEPA 1995d) and 552.3 (USEPA 2003b) for HAAs. All methods will be evaluated as to their potential to be applied and/or adapted for on-line monitoring of the concentrations of THMs and HAAs in water distribution systems.

SIGNIFICANCE OF PROJECT

The practical benefits of the proposed research are the development of improved capabilities for monitoring THMs and HAAs in drinking water. These methods will provide data for operators that will help them better understand the temporal behavior of the THMs and HAAs at the treatment plant and in the distribution system. The methods that are developed will be simple, relatively inexpensive and automated. If feasible, the methods will be based upon the well-documented and understood USEPA Methods for THMs and HAAs in drinking water.

EXPECTED OUTCOMES

A minimum of four different semi-automated methods, two for THMs and two for HAAs, will be developed that are applicable to monitoring THMs and HAAs concentrations in drinking water. The availability of multiple methods based upon different chemistries for monitoring DBPs in drinking water will provide higher quality data on which to base predictive models. These analytical methods will be evaluated initially based upon method detection limits (MDLs), accuracy and precision as defined by the EPA. These studies will be carried out at a minimum of three different THMs concentration levels in the range 0.1 to 100 μg/L. For promising methods, MDL, accuracy and precision studies will be performed for THMs in high ionic strength (~3.5 to 4 mM salt) water and in the waters of at least two different municipalities that record data on THMs.

CHAPTER 2
OVERVIEW OF EXISTING ANALYTICAL METHODS FOR MONITORING THMs IN DRINKING WATER DISTRIBUTION SYSTEMS

ENVIRONMENTAL PROTECTION AGENCY METHODS

There are four well-established analytical methods for measuring THMs concentrations in drinking water. USEPA Method 502.2 (1995a) uses purge and trap (P&T) with GC with a photoionization detector (PID) and electrolytic conductivity (Hall) detector in series. The Hall detector will detect THMs. USEPA Method 524.2 (1995b) uses P&T coupled to GC-MS. USEPA Method 551 (1990a) and USEPA Method 551.1 (1995c) employ liquid-liquid extraction (LLE) of the water sample followed by analysis using GC with an electron capture detector (ECD). Table 2.1 shows the expected MDLs, accuracy and precision estimates for these standard methods for drinking water analysis. The accuracy is measured as mean percent recovery (mean % recovery) and the precision is expressed as percent relative standard deviation (% RSD).

Essentially, all of these are reliable methods, with MDLs on the order of single and tens of parts per trillion (ng/L) for the individual THMs. This is well below the MCL for total THMs of 80 μg/L. The mean % recoveries and % RSD are well within acceptable ranges. The most sensitive method for THMs is clearly USEPA Method 551 (1990a). However, the use of the MS detector with USEPA Method 524.2 (1995b) allows for confirmation of chemical identity and selected ion monitoring (SIM).

There are two basic differences that exist between these methods: (1) The sample preparation/introduction step (either P&T or LLE, etc.) and (2) The detectors that are used (PID, Hall, ECD or MS). The USEPA Methods are very reliable and based upon clever chemistries. They work well for routine measurement of THM concentrations in drinking water when a few samples need to be analyzed over a long time period, for example, monthly and quarterly monitoring schedules. The sample throughput for on-line monitoring is expected to be more demanding, providing data on the minutes to hours time-scale. This data needs to be readily available to operators for immediate decision-making/action. Since the detection devices employed by the standard USEPA methods provide low enough MDL estimates (compared to MCLs for THMs) and suitable accuracy and precision, one alternative would be to retain the analytical methods while automating the sampling-sample introduction process to increase sample throughput. This is the key to on-line monitoring of DBPs like THMs and HAAs.

ALTERNATIVE THMs METHODS APPEARING IN THE LITERATURE

There are numerous methods appearing in the literature for measuring THMs concentrations in drinking water. The methods have been very loosely grouped according to the following categories: GC based methods, GC-MS based methods and colorimetric based methods.

GC Based Methods for THMs

There have been several approaches to sample extraction in GC analysis. Some rely on conventional LLE as a concentration step as do the USEPA Methods. Other techniques such as headspace sampling (HS), solid phase microextraction (SPME), purge and trap-GC (P&T-GC),

and supported capillary membrane sampling-GC (SCMS-GC) have been applied to THMs analysis. Methods adapting these techniques should be readily automated and adapted for on-line monitoring of THMs in drinking water distribution systems.

LLE GC-ECD for THMs

More recently, Nikolaou et al. (2002b) compared four analytical methods, largely derived from the USEPA methods, for measuring THM concentrations in drinking water. Their study compared LLE GC-ECD, LLE GC-MS, P&T GC-MS, and HS GC-MS. For the LLE GC-ECD method, an HP 5890 GC with ^{63}Ni ECD was used. The column was a DB-1 (30 m × 0.32 mm × 0.25 μm f.t.). The injection technique was split with a ratio of 1:25. The carrier gas was He (1.6 mL/min) and the make-up gas was nitrogen (N_2). The injector temperature was 175°C and the detector temperature was 300°C. The oven temperature was 39°C for 12 minutes. The method was evaluated at five different concentration levels covering the range 0.5 to 20 μg/L. The MDLs for $CHCl_3$, $CHBrCl_2$, $CHBr_2Cl$ and $CHBr_3$ were 0.01 μg/L, 0.005 μg/L, 0.007 μg/L, and 0.010 μg/L, respectively. The mean % recoveries for $CHCl_3$, $CHBrCl_2$, $CHBr_2Cl$ and $CHBr_3$ averaged 100%, 100%, 101% and 102%, in that order for five studies done at five different concentration levels. The % RSD for $CHCl_3$, $CHBrCl_2$, $CHBr_2Cl$ and $CHBr_3$ averaged 3%, 5%, 3% and 3%, correspondingly.

The method was accomplished with commercially available instrumentation and derived from USEPA based methods. The sample preparation time is approximately 30 minutes and the chromatographic analysis time is 12 minutes. The sampling rate ranges from 1.2 to 1.5 sample/hr. The operator skill level is estimated at medium level, requiring a trained analyst with experience in chromatographic instrumentation.

P&T GC Methods for THMs

Allonier et al. (2000) proposed a P&T GC-ECD method for the analysis of THMs in seawater. They applied this method for measuring THM concentrations in three different coastal waters. The GC was an HP 6890 equipped with ECD. The column was a narrow-bore HP-Innowax (30 m × 0.25 mm i.d. × 0.25 μm f.t.). The carrier gas was He at 2.5 mL/min. The make-up gas was N_2 at 60 mL/min. The injector temperature was 200°C and the detector temperature was 250°C. The initial oven temperature was 40°C for five minutes, increased to 130°C at 8°C/min, increased to 180°C at 25°C/min and held for five minutes.

The P&T system was HP 7695 model with a glass U-tube sparger without a frit (5 mL). The trap (30.5 cm × 0.312 cm i.d.) was packed with Tenax GC, silica gel and activated carbon. The P&T system was directly coupled to the GC by means of a transfer line and a split valve (split ratio of 1:5). Their results, at a constant desorption time of four minutes, showed that increasing the purge pressure over the range 2.5 bar to 3.5 bar did not significantly improve the purge efficiency. The purge efficiency was slightly improved by increasing the purge time. Ultimately, they chose a purge time of 11 minutes as a compromise between sensitivity and total analysis time. During the desorption step, increasing the temperature of the trap decreased efficiencies by 10% to 20% over the temperature range 180 to 225°C. Increasing the sample temperature from room temperature to 60°C slightly improved the purge efficiency for all compounds except $CHBr_3$. Room temperature was chosen in the final method to simplify analysis. Unexpectedly, a slight decrease in the overall efficiencies was observed for all THMs when the temperature of the transfer line was increased from

100 to 200°C. Salt addition to the sample was used to enhance the purge efficiency of $CHBrCl_2$, $CHBr_2Cl$ and $CHBr_3$.

The MDLs for $CHCl_3$, $CHBrCl_2$, $CHBr_2Cl$ and $CHBr_3$ were 0.05 µg/L, 0.02 µg/L, 0.02 µg/L and 0.07 µg/L, respectively. This method was evaluated at two different concentration ranges. With each THM at 3 µg/L concentration level, the % RSD ranged from 2% to 8%. Mean % recoveries were between 93% and 112%. The correlation coefficients (r) ranged from 0.996 to 0.997. With each THM at the 50 µg/L concentration level, the % RSD ranged from 8% to 9%. The mean recoveries ranged from 98% to 99%. The r values ranged from 0.999 to 0.997. Bromochloromethane and dibromochloropropane were used as internal standards. The method appears to have been done with commercially available instrumentation and is derived from USEPA methods. The average total analysis time for a sample was approximately 50 minutes including a 23 minutes chromatograph run time. Since one sample can be purged while another is being analyzed, the total analysis time is on the order of 30 to 40 minutes. The sample analysis rate would be approximately 1.2 to 1.5 sample/hr. The skill level of the analyst is estimated to be medium-high, requiring experience with P&T and chromatographic instrumentation.

Zygmunt (1996) described a laboratory-built P&T device (Figure 2.1) used to analyze the THMs in standard solutions and real samples. Purge parameters such as purge gas flow-rate and volume; purge temperature and ionic strength were optimized. Stock solutions were prepared for individual THMs in methanol (MeOH). Working standard solutions containing mixtures of the THMs were prepared using the THM stock solutions and "zero water." "Zero water" was 10 MΩ reagent water that had been boiled for 30 minutes. Prior to use, this water was purged with argon (Ar) for 20 minutes.

The P&T system had a macrotrap and a microtrap. The macrotrap consisted of a glass tube with 105 milligrams of Tenax GC and 180 milligrams of Carbosieve III S. The microtrap was a glass tube with 15 milligrams of Tenax GC and 21 milligrams of Carbosieve III S. The traps were packed in the laboratory from commercially available packings. The purge gas was Ar at 30 mL/min and the purge gas volume was 600 mL. The trap temperature during sorption was ambient (~25°C), and the trap temperature during desorption was 250°C. The desorption time from the macrotrap was four minutes. The desorption time from the microtrap was two minutes.

The GC conditions were as follows: The column was a DB-1, (30 m × 0.32 mm i.d. × 5 µm f.t.) dimethylpolysiloxane column. The carrier gas was He at 2.0 mL/min and N_2 was the ECD make-up gas at 70 mL/min. The injector temperature was 175°C. The oven temperature was 50°C for two minutes then increased to 200°C at 10°C/min where it was held for five minutes. The total analysis time was 48 minutes. The purge parameters were optimized including purge gas volume, sample temperature and ionic strength.

Detailed MDL studies were not reported other than citing that they were at the single ng/L level. The total system response was determined for the THM concentration range from 50 ng/L to 1 µg/L. In this range, their values ranged from 0.9918 for $CHBr_3$ to 0.9987 for $CHCl_3$. The system was made with a combination of commercially available and non-commercially available components. The analytical method was derived from USEPA based methods. The analysis time was estimated to be approximately 70 to 80 minutes per sample meaning that and approximate sampling rate of 0.8 sample/hr is possible. The skill level of the analyst is estimated to be high considering that the system was build in house and requires knowledge of P&T and chromatographic instrumentation. The researchers applied the method for THM analysis in a variety of matrices including soft drinks, tap water and in infusion fluid. The principle advantages of the method were the elimination of cryofocusing and the low MDLs on the order of 1 ng/L.

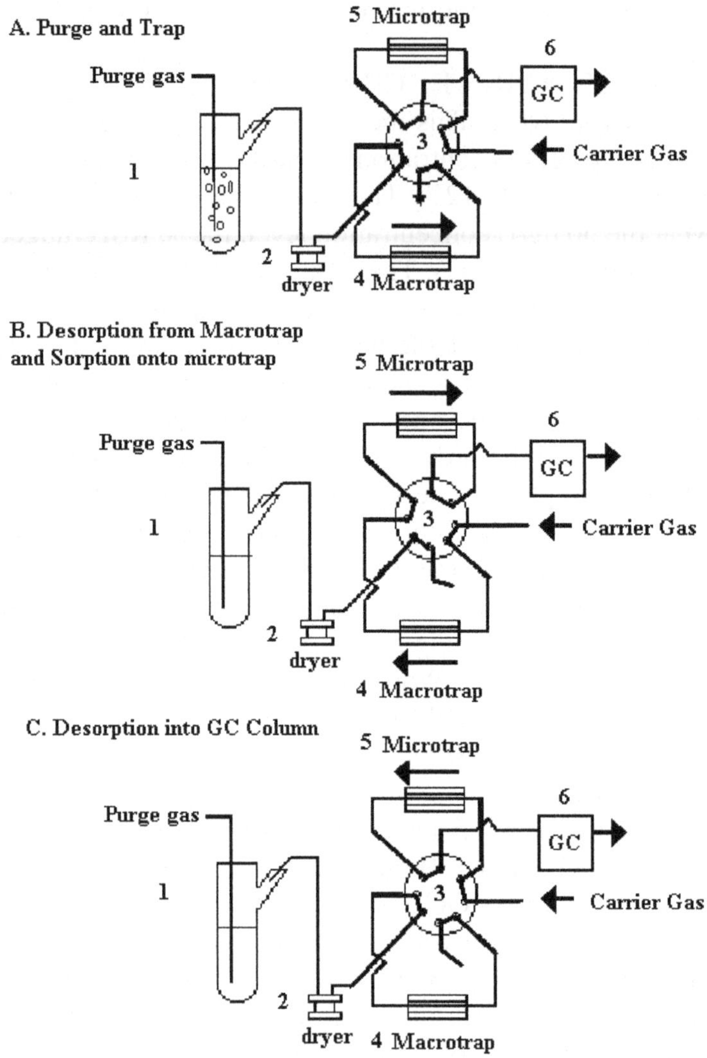

Figure 2.1 Schematic diagram of the laboratory built apparatus for P&T. 1 = purge vessel, 2 = 1.2 meter Nafion dryer embedded (Adapted from Zygmunt 1996).

In the P&T GC-MS method of Nikolaou et al. (2002b), the GC-MS was operated as described previously in their liquid-liquid GC-MS method except that the solvent delay was set to zero. The purge gas was He at 44 mL/min for 11 minutes at 30°C. A 245°C pre-heat temperature was used before desorption at 250°C for three minutes. The bake temperature was 260°C for eight minutes. The transfer line temperature and the valve temperature were 200°C. The moisture control system was set at 120°C.

The method was evaluated at seven different concentration levels covering the range from 0.10 to 40 µg/L. The MDLs for $CHCl_3$, $CHBrCl_2$, $CHBr_2Cl$ and $CHBr_3$ were 0.01 µg/L, 0.05 µg/L, 0.01 µg/L, and 0.01 µg/L, in that order. The mean % recoveries for $CHCl_3$, $CHBrCl_2$, $CHBr_2Cl$ and $CHBr_3$ averaged 99%, 103%, 95% and 94% respectively for five studies done at seven different concentration levels. The % RSD for $CHCl_3$, $CHBrCl_2$, $CHBr_2Cl$ and $CHBr_3$ averaged 5%, 7%, 5% and 5%, respectively.

The method was conducted with commercially available instrumentation and was derived from USEPA based methods. The skill level of the analyst is estimated to be medium-high as it requires knowledge of P&T, chromatographic and MS based methodologies. The analysis time is about 45 to 50 minutes, giving an estimated sampling rate of 1.2 sample/hr.

Chen and Her (2001) developed a continuous-flow P&T GC-MS system for on-line monitoring of THM concentrations in drinking water. The system was constructed from a combination of commercially and non-commercially available components. The analysis was loosely derived from USEPA based methods. Since construction of the device required considerable expertise and requires knowledge of P&T, chromatography and MS, the skill level of the analyst is high. The MDL values for $CHCl_3$, $CHBrCl_2$, $CHBr_2Cl$ and $CHBr_3$ were 10 ng/L, 25 ng/L, 40 ng/L and 50 ng/L, correspondingly. The cycle time could be shortened to five minutes in some cases resulting in a sampling rate of 15 samples/hr. The system was shown to respond well to changes in THM concentrations. Sample carryover was estimated to range from 4.3% to 6.7%. The response lag time was approximately five minutes thus limiting sampling to this interval. The % RSD of samples ranged from 1.4% to 10.5%. The accuracy expressed as % difference between known and measured concentrations ranged from 1.9% to 2.9%. The system was applied in a five-hour monitoring interval at the single μg/L concentration level and showed little variation in the THM concentrations.

Zhang, Qian and Yang (2001) compared P&T GC-MS and P&T GC-ECD for measuring THMs in drinking water. The instrument they used for P&T GC-MS was an HP 6890 GC equipped with HP 5973 MSD, HP P&T auto-sampler and Tekmar 3000 concentrator was used. The trap was Carbopack B/Carboxen 1000 &1001. The GC column was DB 624 (25 m × 0.2 mm i.d. × 1.12 μm f.t.). The MS was set at scan mode (m/z 30 amu to 550 amu in 6.1 scans/second). The interface and ion source temperature were set at 280°C and 230°C, respectively. The following signals were used for quantitation: m/z 82.9 amu, 84.9 amu for $CHCl_3$, 82.9 amu, 84.9 amu for $CHBrCl_2$, 128.8 amu, 126.8 amu for $CHBr_2Cl$ and 172.7 amu, 170.7 amu for $CHBr_3$. For P&T GC-ECD they used an HP 6890 GC equipped with an HP micro-ECD. The sample concentrator was a Tekmar 3100 with a Tenax silica gel/charcoal trap. The GC column was an HP-5 (30 m × 0.2 mm i.d. × 1.12 μm f.t.).

Purge efficiencies were measured the THM species. The efficiency of $CHCl_3$ was 100%. The purge efficiency for $CHBrCl_2$ and $CHBr_2Cl$ were more than 95% but only 75.0% for $CHBr_3$. For the P&T GC-ECD, the MDLs for all THMs were at the sub-ng/L concentration levels. For P&T with GC-MS, the MDL values were at sub-μg/L level. Detailed MDL, accuracy and precision results were not reported.

The method was carried out with commercially available instrumentation and derived from USEPA methods. The skill level of the analyst is medium-high. The analysis time ranged from 30 to 40 minutes giving a sample analysis rate of approximately 1.5 to 2 sample/hr. The method was used to measure THM concentrations in raw water samples from reservoir and river waters in Singapore as well as recreation water from a wastewater treatment plant.

Head-Space GC Based Methods

Wang, Simmons and Deininger (1995) described a simple sparging system to sample THMs in flowing water using a portable GC and a "dynamic HS" system with GC-ECD (Figure 2.2). The system used a simple gas washing bottle modified to allow drinking water to continuously flow through the purge chamber. Air was used to purge the THMs from the continuously

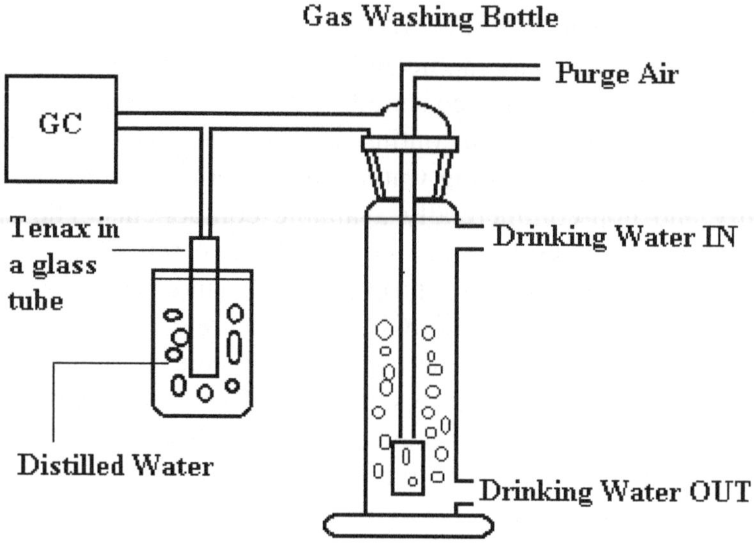

Figure 2.2 Diagram of the dynamic HS sampling device (Adapted from Wang, Simmons and Deininger 1995)

flowing stream. The THMs were trapped on Tenax packed in a glass tube, and introduced into the GC system for analysis.

The MDL for $CHCl_3$ $CHBrCl_2$, $CHBr_2Cl$, and $CHBr_3$ were 0.10 μg/L, 0.01 μg/L, 0.01 μg/L and 0.13 μg/L, respectively. The mean % recoveries for $CHCl_3$ $CHBrCl_2$, $CHBr_2Cl$, and $CHBr_3$ were 104%, 148%, 99% and 85%, sequentially. The % RSD values for $CHCl_3$ $CHBrCl_2$, $CHBr_2Cl$, and $CHBr_3$ were 7.7%, 2.6%, 6.0% and 8.5%, respectively.

The method was carried out with a combination of commercially and non-commercially available materials. The chemical basis of the method was loosely derived from USEPA based methods. The skill level of the analyst is estimated to be medium. The sample analysis time was approximately 30 to 40 minutes per sample. This gives a sample rate of about 1.5 to 2 sample/hr. The strength of this proposed method is that is provides a relatively inexpensive option to traditional P&T instrumentation. It appears to be readily adaptable to on-line monitoring via the built in flow cell by way of a gas washing bottle.

Kuivinen and Johnsson (1999) proposed a HS technique used in combination with GC. The GC was a HP 5890 GC with a ^{63}Ni ECD. The column was a Chrompack CP-SIL 13 CB (25 m × 0.32 mm i.d. × 1.2 μm f.t.). The carrier gas was He (0.72 mL/min to 1.0 mL/min). The make-up gas was N_2 at 50 mL/min. The split flow rate was 30 mL/min. The splitless time was 0.2 minutes. The injector temperature was 200°C and the detector temperature was 280°C. The initial oven temperature was 60°C for three minutes then increased to 70°C at 4°C/min, where it was held for two minutes. The temperature was then increased to 100°C at 4°C/min and held for five minutes.

Separate vials containing samples, blanks and standards were heated at 30°C in a heating block for 30 minutes and shaken for one minute. After two minutes of equilibration, the HS syringe compressed 0.5 mL HS to half its original volume and was injected into the GC. The sample preparation time was 35 minutes and the chromatographic analysis time was 20 minutes giving a total analysis time of 55 minutes.

The limits of quantitation (LOQ) for the method were less than 0.1 μg/L for three of the THM analytes and 0.2 μg/L for $CHBr_3$. The % RSD values were 8% and 0.8% for concentrations of 1 μg/L and 50 μg/L, respectively. The method was evaluated at three different concentration ranges. With each THM at 0.1 μg/L concentration level, the % RSD ranged from 3% to 20%. Mean recoveries were between 89% and 91%. With each THM at the 1 μg/L concentration level, the % RSD ranged from 5.8% to 13.5%. The mean recoveries ranged from 98% to 110%. With each THM at 50 μg/L concentration level, the % RSD ranged from 0.8% to 3.3%. Mean % recoveries ranged from 99% to 103%.

The linear range of the method is between 0.1 μg/L and 75 μg/L, and the r values ranged from 0.996 to 1.000 for each THM. Optimization studies showed that horizontal shaking gave a more rapid equilibrium time than vertical shaking. Increasing the heating time from 30 to 60 minutes showed that there was no significant difference in sensitivity. Increasing the heating block temperature from 30 to 50°C slightly increased the sensitivity.

The method was carried out with commercially available instrumentation. The chromatographic analysis is derived from USEPA based methods. The skill level of the analyst is rated at medium, requiring knowledge of HS and chromatographic analysis. The analysis time for a sample is 53 minutes giving a sampling rate of about 1.1 sample/hr. The advantages of this method were it is relatively easy to use. It is inexpensive and minimizes the use of organic solvents. However, the method was relatively slow compared to membrane introduction mass spectrometry (MIMS).

Cancho, Ventura and Galceran (1999) monitored the behavior of halogenated DBPs including THMs and HAAs in the water treatment plant of Barcelona. THMs were analyzed by HS-GC. The sampling and analysis procedures for THMs was as follows: 8 mL of sample, 4 μL of a methanolic solution of bromochloromethane and 1, 4-dichlorobutane (50 mg/L and 500 mg/L, respectively) were added in 10 mL glass vials. The vials were capped and incubated at 70°C for 15 minutes. The gas phase was injected into the GC-ECD. The preparation time was 25 minutes. For the GC analysis, a Fisons HS 8000 was used for injection. 0.5 mL of gas phase was injected at the split ratio of 1/20 and held at 225°C. The column was a DB-624 (30 m × 0.32 mm × 1.8 μm f.t.). He at 85 kPa and N_2 at 100 kPa were used as the carrier gas and make-up gas, respectively. A ^{63}Ni ECD was operated at 300°C. The oven temperature was initially set at 40°C held for ten minutes and increased to 65°C at 2.5°C/min, increased to 140°C at 9.0°C/min and then at 15°C/min to final temperature 220°C where it was held for ten minutes. Specific details of the MDL, accuracy and precision studies were not reported. The method was carried out with commercially available instrumentation and the chromatography was derived from USEPA based methods. The total time for analysis was estimated at about one hour including a chromatographic time of 34 minutes. The sample analysis rate is expected to be 1 sample/hr.

Batterman et al. (2002) measured THM partition coefficients in blood, urine, water, milk and air. The THMs concentrations were measured using a Tekmar 7000 HS autosampler coupled to a Varian 3700 GC equipped with a linearized ECD. For this study, a sufficient equilibrium time was needed. After adding known quantities of THM into 22 mL sealed vials, the vials were incubated at 37°C for 1.5 hours. Each vial was shaken at low power for two minutes prior to sampling. The injection volume was 1 mL. The column was a packed stainless steel of 20% THEED on 60/90 mesh firebrick (2.0 m × 4 mm i.d). The oven temperature was held at 40°C for two minutes, ramped at 20°C/min to 75°C, and held for one minute. To improve quantification of $CHCl_3$, a separate program with a constant oven temperature of 30°C was used. The MDL values for $CHCl_3$, $CHBrCl_2$, $CHBr_2Cl$, and $CHBr_3$ were 0.1μg/L, 0.03 μg/L, 0.04 μg/L, and 0.5 μg/L, respectively. Details of the accuracy and precision such as mean % recovery and % RSD were not reported.

The method was conducted with commercially available instrumentation and derived from USEPA based chromatographic methods. The sample preparation time was 100 minutes including a relatively short (5 minute) chromatographic time. The sample analysis rate averaged about 0.6 sample/hr.

SPME Based GC Methods for THMs

Solid phase micro extraction (SPME) is a simple, cost-effective, time saving extraction technique developed by Pawliszyn and associates (Sigma Aldrich 2003; see Figure 2.3). The SPME device consists of a syringe-type holder and a 1 cm long fused-silica fiber coated with specific polymer fitted in the syringe needle. After the syringe needle pierces through the septum of a sample vial, the fiber is immersed into either a water sample or the headspace to extract the organic analytes onto the polymer coating. After sampling is completed, the fiber is inserted into a heated GC injector for desorption and analysis. In addition to extracting THMs this technique is also applicable to a wide variety of other organic compounds both in aqueous matrix and in headspace. The SPME sampling devices and a variety of SPME fibers are commercially available from Supelco Inc. Examples of different fibers are shown in Table 2.2. The largest variability in the SPME fibers derives from the different polymer coatings that are used to construct the fiber. There are eight major types of polymer coatings. An additional variable is the thickness of the polymer coating. The four basic types of polymer coatings: polydimethylsiloxane (PDMS), polyacrylate (PA), divinylbenzene (DVB) and carbowax (CW). These four coatings are sometimes combined to form other coating varieties such as polydimethylsiloxane-divinylbenzene (PDMS-DVB), carbowax-divinylbenzene (CW-DVB), divinylbenzene-carboxen-polydimethysiloxane (DVN/carboxen/PDMS), and carboxen-polydimethysiloxane (carboxen/PDMS).

The fibers are classified according to their extraction mode and polarity. The polar fibers include the PA-coated fibers and the CW-DVB coated fibers. The remaining fibers are non-polar or bi-polar. The non-polar fibers have a PDMS coating, and the bi-polar fibers are primarily non-polar, but will extract some polar analytes efficiently. Bare fused silica is listed as an adsorbent type fiber because the surface of the fused silica interacts with the analytes. This fiber is reported mainly for comparison as it is seldom used in practice without a coating. Uncoated fused silica is relatively fragile and thus not typically used.

The other way to classify SPME fibers is by extraction mechanism. In general there are two types of mechanisms, absorption and adsorption. Absorbent fibers extract by partitioning into a liquid-type coating. The analytes are retained by the thickness of the coating and size of the analyte. The polarity of the fiber coating may enhance the attraction of an analyte to a particular coating. There is virtually no competition between analytes. Extraction efficiency at a simple level is basically determined by how fast the analytes migrate in and out of the phase. Conversely, adsorbent type fibers contain porous particles suspended in a liquid phase. The particles retain analytes in the pores or on the surface. Currently, all of the adsorbent type fibers are on a StableFlex core to reduce breakage and increase bonding strength. These fibers extract by physically interacting with the analytes. Classical adsorption mechanism theory suggests that, due to limited adsorption sites, the lighter analytes will be displaced by the heavier analytes as the concentration increases (Sigma Aldrich 2003).

The surface area of DVB is relatively large. The material is primarily mesoporous, with a moderate amount of macropores. There are very small micropores present in this DVB material. CAR-PDMS has a similar surface area as DVB. The major difference between DVB and

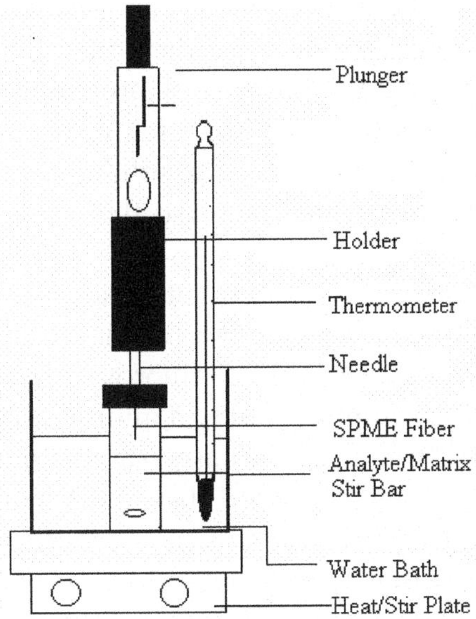

Figure 2.3 Diagram of an SPME sampling assembly (Adapted from Sigma Aldrich 2003)

CAR-PDMS is that there are a much higher percentage of micropores in CAR-PDMS. The pore structure allows analytes to desorb efficiently.

In general, with a 100 µm PDMS fiber, non-polar compounds with high distribution constants give lower MDL values than more polar analytes with lower distribution constants. Methods developed for sampling branched aromatic compounds and chlorinated alkenes appear to exhibit the lowest MDL values. With extremely volatile compounds, the analyte will begin to desorb from the SPME fiber before the sample is injected in to the GC. Therefore, when analyzing volatile compounds, it is best to minimize the time between extraction and desorption and to maintain consistent timing for each step. This strongly suggests that for maximum efficiency, SPME need to be fully automated.

There are commercially available GC inlet liners that are designed for optimum sample introductions for specific injection techniques. SPME is a solventless extraction method, since no solvent is injected and no split is employed, a 0.75 mm i.d. inlet liner can be used. Reduced volume in the 0.75 mm i.d. inlet liner increases the linear velocity through the liner and rapidly introduces the analytes onto the column in a narrow band. To minimize sample loss or peak tailing, the inlet liner must be inert (Sigma Aldrich 2003).

In general, there are several factors that must be controlled for good results using SPME. Agitation, addition of salt, pH adjustment, and immersion of the fiber in the liquid sample improve recovery of difficult-to-extract compounds. There are several factors that can affect sampling in SPME. For quantitative results, the factors must be standardized from analysis to analysis.

The particular sample preparation for SPME may change with different sample matrices since the sample matrix will influence the precision, accuracy and chromatographic results. Extraction time is critical for the sample to establish equilibrium with the SPME fiber coating. Extraction typically takes 15-20 minutes, but can be as short as 30 seconds. Headspace extractions are usually shorter than immersion sampling.

The temperature is typically held constant so that good precision is achieved. Heating the sample during headspace extractions will increase the concentration in the headspace and help improve sensitivity and shorten the extraction time. Sample agitation is important to reduce the equilibrium time and improve the accuracy and precision results. Adjusting the pH or adding salt can improve the extraction efficiency by changing the solubility of the analytes in the sample. The addition of 25%-30% solution of NaCl will increase the ionic strength of the sample, which reduces analyte solubility.

Headspace sampling sensitivity is best when headspace volume is small. The recommended headspace volume is between 30% and 50% of the vial. The position of the fiber should be at the same depth in the headspace every time to improve reproducibility. Immersion sampling sensitivity is improved by filling the sampling vial to a minimum of 80% total volume. Immersion sampling works best for low concentration water based sample matrices.

Once samples are extracted onto the fiber, they must be desorbed into the GC. Splitless injection is generally required with SPME to focus the analytes on the chromatographic column and prevent SPME fiber from bending. For quantitation, standardization is required. External standard calibration compares detector responses from the sample to the responses from the target compounds in the calibration standard. Comparison of the sample extract's detector response to the calibration curve determines the amount of analyte in the unknown sample. Internal standard calibration requires the addition of a known amount of a known compound (the internal standard) into the calibration standards and samples. Internal standardization is used when the sample matrix is complex with gaseous or liquid mixtures or when the chromatographic results vary from run to run. Standard addition is also used with SPME by adding known concentrations of the actual analyte of interest to multiple aliquots of the sample. Standard addition is typically used when it is not possible to prepare a blank matrix. Surrogate and matrix spike compounds are used to assess matrix interferences that might bias quatitation.

SPME has several advantages to many more traditional techniques. As long as carefully designed and calibrated sampling procedures are developed, it is well suited for simple sampling as it is passive, portable and easily adapted for field sampling. Analysis by SPME is virtually solvent free. SPME works well with GC and GC/MS analysis. Cross contamination is minimized. The SPME fibers are reusable up to 100 times and may be applied to a wide array of compounds, meaning a minimal investment in separate SPME fibers could be used for THMs, HAAs or other compounds of interest. SPME is relatively sensitive compared to other sampling techniques.

For THMs analysis, the 100 μm PDMS fiber is used with extraction time generally in the range of 10 to 15 minutes. 25% NaCl salt gives good precision and increases reproducibility. Other parameters like temperature, agitation and fiber height should be kept constant. External standardization is generally used for quantitation. For HAAs, the 75 μm CAR-PDMS has been used with success. The mean micropore diameter of CAR is lower than that of divinylbenzene, so it is ideal for SPME analysis of small molecules in the C_2-C_6 range, such as HAAs methyl esters. For the analysis of HAAs by GC, a prior derivatization step is necessary because of their low volatility and high polarity. The derivatized HAAs are then extracted using SPME and introduced into a GC for separation, detection and quantitation. The required parameters that increase the precision should be kept constant. Internal standardization was used for calibration. Surrogates may also be used to assess matrix interferences.

SPME is an attractive alternative to traditional GC sample introduction techniques such as LLE and P&T. The SPME sampling experiment consists of a syringe-type holder and a 1 cm long fused-silica fiber coated with a polymer coating fitted in the syringe needle. The syringe needle pierces the septum of a sample vial so that the fiber is immersed directly into the water sample (SPME) or the HS above the water sample (HS-SPME) to extract THMs onto the polymer coating of the SPME fiber. Once sampling is completed, the fiber is inserted into a heated GC injector for desorption followed by separation and quantitation. SPME can be applied to sampling THMs in aqueous solution and in HS analysis.

Chen, Mao and Hsu (1997) proposed a method for measuring the concentrations of THMs using SPME coupled with a Hitachi G-3000 GC with a ^{63}Ni ECD. The SPME fiber was a fused silica fiber coated with a 7 μm thick stationary phase of PDMS. The GC column was a DB-5 (30 m × 0.25 mm i.d. × 0.25 μm f.t.). The carrier gas was N_2 at a flow rate of 0.5 mL/min. The make-up gas was N_2 at 45 mL/min. The injector temperature was 200°C and the detector temperature was 250°C. The oven temperature was programmed as follows: The temperature was held at 35°C for three minutes, then increased to 80°C at 10°C/min where it was held for ten minutes. The temperature held at 80°C for four minutes then increased to 120°C at 10°C/min with a final hold time of two minutes. The GC run time was 24 minutes. The total analysis time was approximately 45 minutes.

The SPME procedure was as follows: Sample (2.00 mL) was transferred to a 4 mL sample vials. The SPME fiber was placed in the solution and allowed to equilibrate. The syringe needle was immediately inserted into the GC injector for desorption. The equilibrium time and the desorption time would be optimized beforehand. Oven cryofocusing was used to sharpen the injection band. The cryofocus temperature was set 80°C-100°C below the boiling point of the analyte. The THM concentrations ranged from 1 to 100 μg/L for the study. The % RSD estimates for $CHCl_3$, $CHBrCl_2$, $CHBr_2Cl$ and $CHBr_3$ were 4.0%, 3.0%, 3.6% and 3.9%, sequentially. The mean % recoveries ranged from 112% to 121%.

The method was carried out with commercially available devices and instrumentation. While SPME methods are nor USEPA approved, the chromatographic methods was derived from USEPA based methods. The skill level of the analyst is estimated to be medium, requiring knowledge of SPME and chromatographic instrumentation. The time required for analysis is about 30 to 35 minutes giving an estimated sampling rate of about 1.7 to 2 sample/hr.

Cho, Kong and Oh (2003) reported a method for analysis of THMs in drinking water using a headspace SPME technique (HS-SPME) with an 85 μm CAR/PDMS fiber. Several experimental parameters were investigated including the type of SPME fiber used, the volume ratio of sample to HS, the addition of salts, magnetic stirring, extraction temperature, extraction time and desorption time on the analysis. In addition, the analytical parameters such as linearity, repeatability and MDLs were estimated.

An HP 6890 GC with a ^{63}Ni ECD was used, and the GC conditions were as follows: The column was an HP-5MS (30 m × 0.250 mm i.d. × 1.0 μm f.t.) was used. The carrier gas was N_2 at 1.0 mL/min. The make-up gas was N_2 at 60 mL/min. The injector temperature was 250°C and the detector temperature was 280°C. The oven temperature was 50°C for two minutes, 5°C/min to 110°C, 10°C/min to 250°C and held for two minutes giving a total run time of 30 minutes.

The standard solutions were prepared from a commercially available stock solution containing each THM. Calibration standards were prepared over the concentration range 0.05 to 80 μg/L using reagent water. The internal standard was 1,4-dichlorobutane. Practically, the procedure was to suspend the fiber in the HS of the vial. The temperature was maintained at 35±1°C. After a 30 minutes extraction time, the fiber was retracted back into the needle and transferred to the injection port of GC. The desorption and conditioning times were four minutes. The desorption temperature was 250°C.

Of six fiber types studied, the 85 μm CAR/PDMS fiber worked best for the extraction of THMs. The effect of HS volume was investigated by increasing the sample volume size from 20 to 40 mL (using 60 mL vials) thus changing the HS volume from 40 mL to 20 mL. The results showed that HS volume should be minimized to achieve high-sensitivity HS extraction. The addition of salt changed the ionic strength, which may affect the vapor/liquid equilibrium of extracting membrane. With all comparison studies, all other conditions were held constant with only one parameter varied. Optimizing sample stirring and salt addition were the two variables that improved performance. The mean % recoveries in the standards that had salt added were approximately two times better compared to standards prepared without salt. Recoveries of THMs were 2.5 to 4 times better for standards that were stirred compared to the recoveries with static (non-stirred) standards.

The effect of extraction temperature was more complex. The temperature was varied over the range 20 to 80°C. In some cases, higher temperatures may increase vapor pressure of volatile analytes in the HS—but at the same time reduce the SPME fiber coating-HS (air) partition coefficients. Under non-equilibrium conditions, increasing the temperature can give more rapid transfer of analytes from liquid to vapor phase. The extraction time was increased from 5 to 120 minutes at constant temperature (35°C), and the optimum equilibrium time for $CHCl_3$, $CHBrCl_2$, $CHBr_2Cl$, $CHBr_3$ was 30 minutes. After 120 minutes, the extraction reached equilibrium. The desorption time was varied from 0.5 minutes to four minutes at 250°C. After one minute at 250°C, desorption was complete for each of the THMs.

The linear range of the HS-SPME method was evaluated by preparing calibration curves of peak area relative to the area of the internal standard 1,4-dichlorobutane versus the concentration of each analyte. In practice, a plot of ratio of analyte peak area to internal standard peak area versus the concentration of each compound was prepared. Studies were conducted at three different concentration ranges: 0.05 to 40 μg/L, 0.05 to 60 μg/L, 0.05 to 80 μg/L. Calibration curves were linear over the entire range even though quadratic fits to the calibration curve gave better results over the broadest calibration range. The MDLs were estimated to be in the 0.005 to 0.01 μg/L concentration range. The precision expressed as % RSD ranged from 0.8% to 6.2%.

The methods were carried out with commercially available equipment and instrumentation. The skill level of the analyst is estimated to be medium. The time required for analysis of one sample is about 60 to 65 minutes giving a sampling rate of approximately 1 sample/hr.

A Simple Membrane Sampling Device for Volatiles in Water

Blanchard and Hardy (1984) proposed a method that used a silicone polycarbonate membrane and a collection medium of activated charcoal as a sampling device followed by the quantitative measurement by GC. In the initial membrane apparatus, one side of the membrane was exposed to an aqueous solution and the other side had a stream of N_2 flowing past it. The cell was designed such that the volume of either side was 1 mL and the exposed membrane area was 0.8 cm^2. The sampling devices were composed of 17 mm i.d. glass tubing, 50 mm in length to which a silicone polycarbonate membrane (0.025 mm thick) was affixed with silicone rubber cement. The exposure chambers were 4-L Erlenmeyer flasks in which the devices would be suspended, and magnetic stir bars were used to assure adequate circulation of the test solutions. Aluminum foil was used to cover the rubber stopper used to seal the container to prevent sample losses An HP 5730A dual column GC equipped with dual flame ionization detectors was used for all GC analysis. The packed GC columns were 1% SP-1000 on 60/80 Carbopack B (6 ft. × 1/8 in.) stainless steel columns. The oven temperature was initially set to 100°C for four minutes, and then increased to 220°C at 8°C/min. The total chromatographic run time was 19 minutes.

The results of the initial membrane evaluation showed that the rate of permeation through the membrane was a function of exposure solution concentration over a range of 1 to 50 mg/L for each species. The response time was considered to be the time required to reach 90% response at a given concentration and was found to be less than five minutes in all cases. Increasing the temperature increases the permeation rate constant of each compound; for $CHCl_3$ the increase was 2.86% per increase in °C. The aim of this paper was to evaluate the performance of the membrane sampling device for the analysis of volatile priority pollutants including $CHCl_3$. Linearity was observed from μg/L concentration levels to mg/L concentration levels. The major advantage was that it could provide the time-weighted-average concentration values without electrical power. It is simple and inexpensive with a relatively short response time, approximately five minutes, and total analysis time of 24 minutes. The disadvantage was that the response is temperature dependent.

The membrane sampling method described above was carried out with a combination of commercially and non-commercially available materials. The chromatographic analysis was not derived from USEPA based methods but did use a commercially available gas chromatograph with flame ionization detectors. The analysis time was 25 minutes per sample. The sample analysis rate was about 2.5 sample/hr.

Supported Capillary Membrane Sampling-GC

Duty (2000) and Liao (2001) have explored measurements of THM concentrations in water using a commercially available supported capillary membrane sampling (SCMS) probe as a sampling device for GC. The SCMS probe (Figure 2.4) is comprised of a silicon rubber capillary membrane wound around a metal body that is approximately five inches in length. The capillary membrane is connected to inlet and outlet fittings.

The portion of the SCMS probe containing the wrapped membrane is placed directly in the water to be analyzed. The SCMS-probe can be assembled in a calibration configuration or in a fast-loop/flow cell configuration for analysis of THM concentrations in real time. In the calibration configuration (Figure 2.5), the SCMS probe is immersed in the water sample contained in the calibration vessel and connected to N_2 as the SCMS carrier gas. The outer wall of the capillary membrane is in direct contact with the water sample containing THMs.

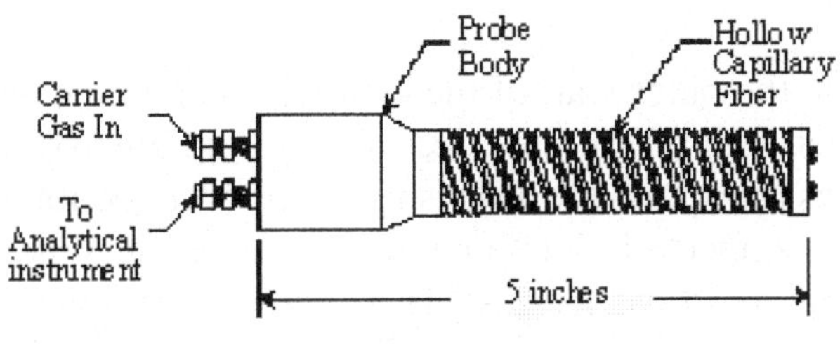

Figure 2.4 The SCMS sampling probe (Adapted from Global FIA 1999)

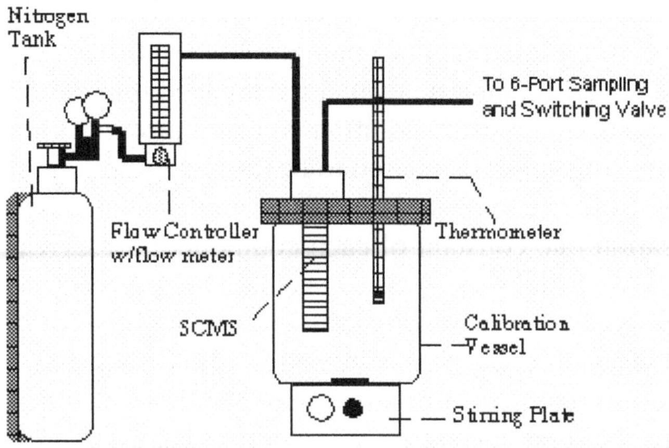

Figure 2.5 The SCMS probe and calibration vessel in the calibration configuration (Adapted from Duty 1999)

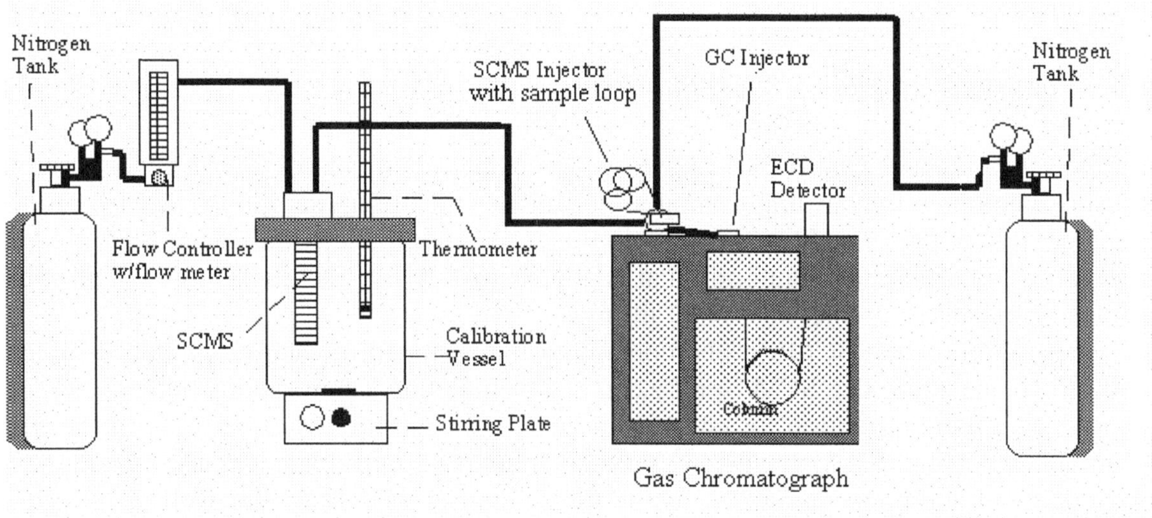

Figure 2.6 The SCMS-GC analyzer. The SCMS-GC is equipped with FID, ECD and PDPID detectors. The configuration shown is SCMS-ECD (Adapted from Duty 2000).

The THMs permeate the outer wall of the capillary membrane and move through the capillary membrane to the inner wall where they desorb from the inner-wall into the flowing carrier stream of N_2. The carrier gas transports the THMs to a six-port injection valve fitted with a sample loop. The injection valve is connected to an analytical instrument. When the injection valve is in the “Load” position, the mixture of carrier gas/THMs passes through a sample loop and is vented to waste. When the injection valve is switched to the “Inject” position, the volume of sample in the sample loop is presented to the analytical instrument. Using such a configuration, a supported capillary membrane sampler-GC (SCMS-GC) was constructed (Figure 2.6) equipped with an FID, ECD and a pulsed discharge photoionization detector (PDPID).

The effectiveness of the permeation process depends on: (1) The mass transport of the sample molecules to the membrane outer surface; (2) The solubility of the sample molecules in the polymer; (3) The vapor pressures of the analyte involved; and (4) The temperature (Nilsson et al. 1995). The mass transport rate and vapor pressures of the analytes are strongly affected by temperature. This means that the SCMS probe must be calibrated to achieve quantitative results. This is accomplished by adjusting the SCMS carrier flow rate to a point where the internal flow velocity becomes low enough to allow the analyte to reach vapor-liquid equilibrium with the sample solution. At this point, it is theoretically possible to use Henry's Law and Antoine's equation (Dean 1985) to correct for temperature changes in the sample matrix. Henry's Law indicates that the solubility of a gas is directly proportional to the partial pressure exerted by the gas. Henry's Law constant for $CHCl_3$ and other compounds can be empirically calculated as described by Nicholson, Maguire and Bursill (1984). Antoine's Equation describes the partial pressure of a pure compound over its liquid form as a function of temperature. Antoine's constants can be found in the literature (Kuivinen and Johnsson, 1999) for the analytes of interest at different temperatures. The partial pressure of the analyte can be calculated at a particular temperature using Antoine's Equation and this value substituted into the Henry's Law equation. Thus it is possible to calculate the mole fraction of the analyte.

It is possible to use the aforementioned discussion to correct for temperature variation. In practice, the temperature did not vary significantly (~±1°C) from calibration to calibration. A major accomplishment of preliminary studies was to develop calibration protocols consistent with USEPA recommendations for estimating MDL, accuracy and precision of the methods. These protocols coupled with the design of the SCMS make it readily adaptable to real-time monitoring of THMs in drinking water. The SCMS probe can be immersed into a calibration vessel for batch calibration and sample measurements. The SCMS probe can be inserted directly in process piping or used in an inexpensive, commercially available "flow-cell/fast-loop" adapter (Figure 2.7).

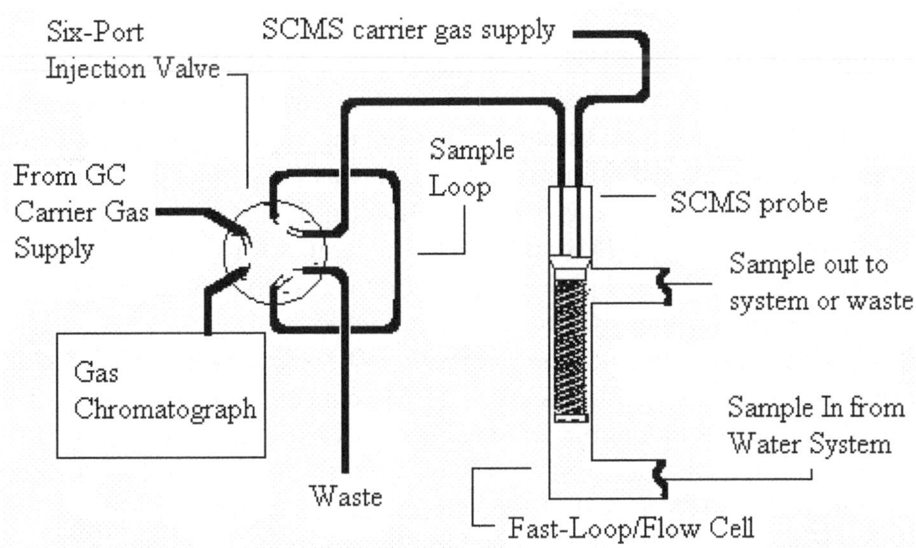

Figure 2.7 Fast-loop/flow-cell configuration (Adapted from Global FIA 1999)

Compared to the traditional sample introduction techniques for GC, SCMS offers several advantages. Preconcentration of analyte is not required. Sample losses or contamination are largely reduced. The SCMS method eliminates the need for extraction solvents. The cost of an SCMS-probe and maintenance is significantly less than other comparable sample introduction devices such as P&T, and the SCMS probe is easier to automate. The automation depends only on controlling a two-position six-port injection valve, an "in-house" software program, Valve-Pro, has been developed and is available as shareware that can easily control valve loading and injection. The SCMS probe can be immersed into the calibration vessel for batch measurements, used in a flow-cell/fast-loop arrangement for on-line monitoring in a distribution system, or inserted directly into process line in the distribution system. It is simple to interface the SCMS to a GC. This means that the well-established USEPA Methods could be readily adapted for real time monitoring of THMs in drinking water distribution systems using the SCMS-GC approach.

Two SCMS-GC methods have been developed for measuring the concentrations of THMs in water. These methods use two different chromatographic columns and two different detectors (ECD or PDPID).

SCMS-GC-ECD Method for THMs. An SCMS-GC-ECD method was developed to measure THMs concentrations in water (Duty 2000). The SCMS probe was interfaced to a Varian 3400 GC with ECD. The optimized flow rate of carrier gas for the SCMS-probe was 5 mL/min. The column was a DB-5 (30 m × 0.32 mm i.d. × 0.25 μm f.t.). The GC carrier gas was N_2 at 90 cm/s linear velocity. The injector temperature was 175°C. The detector temperature was 300°C. The initial oven temperature was 35°C for four minutes. The oven temperature was increased from 35 to 75°C at a rate of 10°C/min. The rate was then increased from 75 to 200°C at a rate of 50°C/min where it was held for three minutes. The injection loop volume was 100 μL.

THMs standards were prepared and calibration plots for the SCMS-GC-ECD method were constructed by plotting the height of each THM peak from the chromatogram as a function of concentration. The MDL for each THM component was calculated using USEPA analysis of variance (ANOVA) procedure (Glaser and Forest 1981; USEPA 2003a). The precision and accuracy were calculated based on USEPA recommendations (USEPA 1996). The total run time was about 14 minutes.

The results of these studies show that the SCMS-GC-ECD method for monitoring THM concentrations in water is a viable alternative to currently available methods. Additional studies have shown that the calibration curves are linear over the concentrations range 0.1 to100 μg/L. The error on the slope of each calibration curve was always less than 10%. The average % error on the slope of each calibration plot was 5%. The MDL (μg/L) for each THM component was less than 1 μg/L for three of the four THMs and promising results for the $CHBr_3$ (6 μg/L). The MDL for total THMs was calculated at 6.9 μg/L. The % RSD is less than 5%. The mean % recovery for standard solutions was greater than 100%, but reproducible.

SCMS-GC-Pulsed Discharge Photoionization Detector (SCMS-GC-PDPID). An SCMS-GC-PDPID method was developed to measure THMs concentrations in water (Liao 2001). The PDPID was developed in the laboratory of Wentworth (Vasnin and Wentworth 1992; Wentworth and Limero, 1989; Wentworth and D'Sa 1992; Wentworth and Cai 1994; Mendonca and Wentworth 1996; Wentworth and Watanesk 1996; Gremaud and Wentworth 1996; Cai, Stearns and Wentworth 1998; Wentworth et al. 1996). A stable low power electrical pulse in helium is used as the source. Compounds eluting from the GC column were ionized by photons from the discharge zone. Resulting electrons were focused toward the collector electrode by the two bias electrodes. The principle mode of ionization was photoionization by radiation arising from the transition of diatomic helium (He_2) to the dissociative ground state known as the Hopfield emission (energy range from the He_2 is 13.5 eV to 17.7 eV).

The PDPID can be operated in several modes to achieve nearly universal detection. For example, in He ionization mode (PDHID), the detector is essentially non-destructive, meaning it could be assembled in-line with a MS, and highly sensitive. The response to organic compounds is linear over five orders of magnitude with minimum detectable quantity (MDQ) values in the low or sub picogram range. The response to fixed gases is positive with MDQ values in the low μg/L range. The PDHID response is universal except for neon, which has an ionization potential of 21.56 eV. Additional modes of operation were obtained by changing the discharge gas from 99.9999% He to He doped with argon, krypton, or xenon. This changed the discharge emission profile, resulting in resonance atomic and diatomic emissions of the rare gas added. These detectors have been called PDPID with the dopant gas used as a prefix, (example: Ar-PDPID).

A very useful mode of operation of the PDPID is in electron capture mode (PDECD). The PDECD is very selective for halogenated organics like the ECD. However, unlike the conventional ECD, the PDPID does not have a radioactive source. The MDQ for halogenated organics is reported to be at the femtogram concentration level. Response characteristics and sensitivities are similar to those of a radioactive ^{63}Ni ECD.

A SCMS was interfaced to a Varian 3380 GC with PDECD. The optimized flow rate for the SCMS-probe was 5 mL/min. The GC column was a VB-1 (30 m × 0.32 mm i.d. × 1.0 μm f.t.). The GC carrier gas was 99.9999% He at 50 cm/s linear velocity. The six-port injection valve was connected directly to the GC column. The detector temperature was 250°C. The dopant gas was 3% xenon in helium. The initial oven temperature was 50°C for four minutes. The oven temperature was increased from 50 to 100°C at a rate of 10°C/min and increased to 200°C/min at 50°C and held for two minutes. The injection loop volume was 100 μL.

The results of preliminary MDL, accuracy and precision estimates have been conducted. In summary, THMs standards were prepared and calibration plots for the SCMS-GC-PD-ECD method were constructed by plotting the height of each THM peak from the chromatogram as a function of concentration. The MDL for each THM component was calculated using USEPA ANOVA procedure (Glaser and Forest 1981; USEPA 2003a), and the method used to calculate precision and accuracy were based on USEPA recommendations (USEPA 1996).

The results of these studies showed that the SCMS-GC with PDECD gives comparable results to the SCMS-GC-ECD for measuring THMs concentrations in water. Calibration curves are linear over the concentrations range 1 to 100 μg/L. The error on the slope of each calibration curve was always less than 10%. The average % error on the slope of each calibration plot was 5%. The MDL (μg/L) for each THM component is less than 1 μg/L for three of the four THMs and improved results for the $CHBr_3$ (1.3 g/L). The MDL for combined THMs was calculated at 2.6 μg/L. The % RSD is less than 4%. The mean % recovery for standard solutions is ~105% and very reproducible.

A comparison of the MDL, accuracy and precision studies (Table 2.3) shows that two different viable SCMS-GC methods were developed for measuring THM concentrations in water. The two methods give similar results. Both are readily automated and could be applied to monitoring THMs at the concentration levels experienced by many water utilities (~ 0 to 100 μg/L and higher).

Each of the SCMS-GC based methods were done with commercially available materials and instrumentation. The chromatographic methods were derived from USEPA based methods. The skill level for the ECD based method is medium and medium-high for the PDPID based method. A typical analysis time in each case was approximately 30 minutes. This gives a sampling rate estimated as 2 sample/hr.

Direct Aqueous Injection-GC-ECD

Biziulk et al. (1996) surveyed the different classes of organic pollutants, including THMs, in ground water and drinking water in the Gdansk district. They used direct aqueous injection of samples into a GC-ECD. The GC was a Carlo-Erba 6180 GC with an ECD. The column was a DB-1 (30 m × 0.32 mm i.d. × 5 μm f.t.). The carrier gas was hydrogen (H_2) at 0.4 m/s. The make-up gas was 99.999% N_2. The detector temperature was 350°C. The injection mode consisted of a 102°C isothermal injection system, cold on column with secondary cooling. The injection volume was 2 μL. The MDLs were 0.01 μg/L, 0.01 μg/L, and 0.02 μg/L for $CHCl_3$, $CHBr_2Cl$ and total-THMs, correspondingly.

This method was done with commercially available instrumentation but was not based on USEPA methodology. The skill level of the analyst is estimated to be low to medium as there are no major sample preparation steps. There were no details provided pertaining to the analysis time and sampling rate.

GC-MS Based Methods

There are several techniques that have been applied to the sample introduction/preparation process. LLE GC-MS has been used along with P&T GC-MS. HS analysis as well as SPME and HS-SPME techniques have been applied. Membrane introduction MS (MIMS) offers many advantages for on-line monitoring of THMs in drinking water.

LLE-GC-MS

Nikolaou et al. (2002b) examined an LLE-GC-MS method for THMs. For this method, an HP 5890 GC with 5971 MS was used. The column was an HP-VOC (60 m × 0.32 mm × 1.8 μm f.t.). The injection technique was split with a ratio of 1:25. The carrier gas was He at 1.3 mL/min. The injector temperature was 175°C and the MS temperature was 280°C. The solvent delay was 14 minutes.

The acquisition mode was selected ion monitoring (SIM) using the following m/z values: $CHCl_3$ (83 amu and 85 amu), $CHBrCl_2$ (83 amu and 85 amu), $CHBr_2Cl$ (129 amu and 127 amu) and $CHBr_3$ (173 amu and 254 amu). The dwell time was 100 ms. The oven temperature was 35°C for nine minutes, increased to 40°C for three minutes at 1°C/min, increased to 150°C at 6°C/min. The method was evaluated at five different concentration levels covering the range from 1 to 50 μg/L. The MDLs for $CHCl_3$, $CHBrCl_2$, $CHBr_2Cl$ and $CHBr_3$ were 0.01 μg/L, 0.02 μg/L, 0.02 μg/L, and 0.03 μg/L, sequentially. The mean % recoveries for $CHCl_3$, $CHBrCl_2$, $CHBr_2Cl$ and $CHBr_3$ averaged 115%, 115%, 109% and 112%, respectively for five studies done at five different concentration levels. The % RSD for $CHCl_3$, $CHBrCl_2$, $CHBr_2Cl$ and $CHBr_3$ averaged 11%, 12%, 9% and 13%, respectively.

The method was carried out with commercially available instrumentation and the chromatographic analysis derived loosely from USEPA based methods. The skill level of the analyst is estimated as medium-high, requiring knowledge of chromatographic instrumentation and MS techniques. The analysis time for one sample is about 65 to 75 minutes including a chromatographic run time of about 35 minutes. This gives a sample analysis rate in the range 0.8 to 0.9 sample/hr.

P&T GC-MS

Gelover et al. (2000) used a GC-MS system equipped with a P&T sampler to survey the water samples from municipal sources and specific sites in the distribution systems in five Mexican cities. The P&T GC-MS was a HP 5890 GC equipped with an HP 5971 MS in SIM mode. The column was 5% phenylmethyl silicone (30 m × 0.22 mm × 0.3 μm f.t.). The carrier gas was He. The injection temperature and detector temperature were set at 100°C and 270°C. The oven temperature was 37°C for 30 minutes, increased to 47°C at the rate of 2°C/min, increased to 80°C at the rate of 5°C/min and finally to 180°C at 15°C/min. The run time was 50 minutes. Detailed MDL, accuracy and precision data was not reported.

The method was carried out with commercially available instrumentation and was largely derived from USEPA based methods. The skill level of the analyst is expected to be medium-high, requiring a working knowledge of P&T techniques, chromatography and MS techniques. The time for analysis of one sample is 50 minutes. This gives an estimated sampling rate of 1.2 sample/hr.

HS GC-MS Method

Takahashi et al. (2003) used HS-GC-MS for investigating the formation of THMs from 37 organohalogen compounds in aqueous solution upon heating. The GC-MS system was an HP 5890 GC, an HP 5972 MS and an HP 7694 HS sampler. The column was a DB 1301 (60 m × 0.25 mm i.d. × 0.25 μm f.t.). The oven temperature was 40°C for seven minutes, increased to 180°C at 5°C/min, increased at 15°C/min to 250°C, where it was held for one minute. The carrier gas was He at 45 psi for one minute and decreased to 15 psi at 99 psi/min. A split ratio of 1:4 was used. The injector temperature was 200°C. The interface temperature was 260°C. The total run time was 41 minutes with 35 minutes sample preparation. The MS ion source temperature was 180°C. $CHCl_3$ and $CHBrCl_2$ were monitored at m/z values 83 amu, 85 amu; $CHBr_2Cl$ was monitored at 129 amu, 127 amu; $CHBr_3$ was monitored at 173 amu, 171 amu, and 4-bromofluorobenzene was monitored at 174 amu. The researchers report that the method is not appropriate for measuring the concentrations of THMs in water. Their results noted that the HS GC-MS method over-estimates the concentration of THMs because THMs are produced from other compounds during the sample introduction process.

The method was carried out with commercially available materials and instrumentation. The chromatographic analysis was derived from USEPA methodology. The skill level of the analyst is rated at medium-high, requiring knowledge of HS based techniques, chromatographic analysis and MS techniques. A typical analysis requires about 75 to 80 minutes giving a estimated sampling rate of 0.8 sample/hr.

In the HS-GC-MS method of Nikolaou et al. (2002b), the GC-MS was operated as described previously in the LLE-GC-MS method except that the solvent delay was set to zero. The 10 mL vials containing 8 mL sample were placed in a water bath at 45°C for 40 minutes. A 500 μL aliquot of the gas phase was withdrawn by a gas-tight syringe and injected into the GC. The GC conditions were the same as the LLE-GC-MS method. The method was evaluated at five different concentration levels covering the range from 0.5 to 10 μg/L. The MDLs for $CHCl_3$, $CHBrCl_2$, $CHBr_2Cl$ and $CHBr_3$ were 0.20 μg/L, 0.05 μg/L, 0.05 μg/L, and 0.10 μg/L, respectively. The mean % recoveries for $CHCl_3$, $CHBrCl_2$, $CHBr_2Cl$ and $CHBr_3$ averaged 88%, 98%, 98% and 99% respectively for five studies done at five different concentration levels. The % RSD for

$CHCl_3$, $CHBrCl_2$, $CHBr_2Cl$ and $CHBr_3$ averaged 17%, 14%, 12% and 15%, respectively. The GC run time was 35 minutes.

This method employed commercially available instrumentation with chromatography conditions similar to USEPA methods. The skill level of the analyst is rated at medium-high as the technique requires knowledge of HS, chromatographic analysis and MS. A typical analysis time required 65 minutes. The sampling rate is in the range 0.7 to 0.9 sample/hr.

Membrane Introduction Techniques With MS

The MS is perhaps the most versatile GC detector, and much of the research with capillary membrane sampling techniques has focused on sample introduction into a MS. The research of Westover and Tou (1974), Cooks and coworkers (Brodbelt and Cooks 1987, Kotiaho et al. 1991, Lauritsen et al. 1992, Wong and Cooks 1995), Budde and coworkers (Slivon et al. 1991, Budde 2001), Pawliszyn and coworkers (Pratt and Pawliszyn 1992a, 1992b), and others have led to many successful applications of this versatile technique. On-line monitoring of a wide array of systems has been possible including environmental analysis and biomedical applications with in vivo or fermentation broth monitoring. MIMS was first used in the late 1970s. The principle was that the water sample to be analyzed permeated the membrane into the vacuum side and then diffused into the ion source of the MS. The advantages of MIMS were fast analysis, low determination limit and low matrix effects, and improved sensitivity relative to static HS or P&T techniques. The disadvantages were: (1) some overlapping fragment ions which were difficult to identify and quantitate due to poor resolution prior to entering the MS, (2) slower (lagging) instrument response time compared to capillary inlets, (3) temperature dependence of the permeation process of the THMs through the membrane and the problems caused by membrane rupture during the use.

Lopez-Avila et al. (1999) developed an automated MIMS system to measure the concentrations of THMs. The automated MIMS system consisted of an HP 5973 MS with turbomolecular pump interfaced to FIA-MIMS autosampler. Specifically the system consisted of (1) a capillary membrane probe (silicon membrane: 0.6 mm i.d. × 1.2 mm o.d × 3.2 cm length; s/p medical grade tubing), (2) a direct insertion probe (DIP) inlet transfer line (304 stainless steel mounted on the HP 5973 MS), (3) a flow injection module to allow injection of liquid samples and to control the membrane probe temperature, and (4) an autosampler handling up to 56, 40 mL vials to deliver samples to the capillary probe controlled by a Windows-based software.

The MIMS probe temperature was 70°C. The sample flow rate was 1.1 mL/min. The ion source temperature was 160°C. The quadrupole temperature was 130°C. The sample loading time was 60 seconds with a ten second rinse time. The injection time was 180 seconds. The mass range scanned was 47-250 amu (full scan); m/z was 83 amu, 129 amu, 173 amu (a typical SIM mode for THMs). The SIM dwell time was 800 ms. The emission current was 35 μA, and the electron voltage was 1260 V.

In practice, the working procedures of the MIMS automated system were as follows. The autosampler needle pierced the septum of a sample vial. The sample vial was pressurized through a small hole in the upper part of the needle. Sample was forced into another small hole at the lower part of the needle. Approximately 10 mL of sample was purged through the transfer lines. After a one minute purge time, 1 mL of sample was delivered by the flow injection module to the capillary membrane. The sample diffused from the membrane directly into the ion source of MS for analysis. The researchers noted that the MDLs for the automated MIMS system depended upon the physiochemical properties of each analyte and the MS response. Overall, the full scan

MDLs were on the order of single μg/L concentration range. The calibration curves for THMs were linear over two orders of magnitude. There were no detailed reports of accuracy and precision of the analytical method other than an estimated % RSD for $CHCl_3$ of 5.7%.

The method was carried out with commercially available instrumentation. It was not based on USEPA methods. The skill level of the analyst is high, requiring knowledge of MIMS. The analysis time is about 5 to 6 minutes per sample giving a sample analysis rate of 10-12 per hour.

Chang and Her (2000) proposed a method for on-line monitoring of THMs in drinking water using a derivative of MIMS called continuous MI-fast GC-MS. Fast-GC refers to the shortening of the GC column which results in a shortening of the run time and an overall loss of resolution. The technique for on-line monitoring of THMs in chlorinated water was MI-fast GC-MS. This combines on-line monitoring characteristics of membrane introduction, rapid GC separation and MS identification analysis.

Experimentally, the monitoring system was comprised of a laboratory-built purge-type membrane introduction system, a cryofocusing unit of a Tekmar AERO Trap Desorber and a Fisons GC 8000/800 GC-MS. The cryofocusing unit was placed between the membrane and the GC-MS. The specific operating parameters of the GC-MS were as follows: The column was a DB-5MS (5 m × 0.250 mm i.d. × 0.25 μm f.t.). Nine cm of the column length was mounted in the cryofocusing unit and trap. The scan rate was 0.112 s/cycle. The column temperature was 50°C and the interface and ion source temperature was 200°C. The base peaks of $CHCl_3$, $CHBrCl_2$, toluene-d8 (internal standard), $CHBr_2Cl$ and $CHBr_3$, at m/z values 83 amu, 83 amu, 98 amu, 129 amu, 173 amu, correspondingly were monitored.

A 4 cm piece of stainless steel tube (0.85 i.d. × 1/8 inch o.d.) and an 8 cm Dow-Corning silastic hollow membrane were used in the MI system. The membrane was mounted inside the stainless tube with the exposed length of 5cm. Carrier gas flowed over the outside of the membrane, while the sample flowed inside the membrane. A "T-valve" connected the sample solution, internal standard and mixing coil together. The internal standard was pumped continuously via a syringe pump and mixed with the sample solution in the mixing coil. This mixture then was transported to the hollow fiber membrane. The carrier gas was He at a flow rate of 1.6 mL/min. During the experiment, a sample stream continuously flowed through the membrane introduction system using a peristaltic pump operating at a flow rate of 1.2 mL/min.

There are three steps in the analysis: prevaporation, preconcentration, and fast GC-MS. During the prevaporation step, THMs were vaporized from the sample stream by permeation across the membrane. The preconcentration step involves changing the temperature of the trap so that THMs are trapped and released into the separation column for fast GC. During this step, the trap temperature was cooled to −165°C using a continuous flow of liquid N_2. It took approximately one minute to cool the trap from 200 to −165°C.

The permeation of water across the membrane produces significant analytical effects. A Tekmar 6000 moisture control system was originally used to minimize the effects of the water but this added several minutes to the analysis time and decreased the recovery of $CHBr_3$. As a result the Tekmar 6000 was removed from the system. The permeation of water decreased sensitivity and precision of $CHBr_2Cl$ and $CHBr_3$ analysis. The amount of water entering the MS caused the baseline to increase slightly resulting in a broad background peak and suppressing the ionization efficiencies of the $CHBr_2Cl$ and $CHBr_3$. This was remedied by using programmed temperature vaporization (PTV) during sample injection. PTV refers to programming the temperature of the trap and the injection time such that water will remain in the trap during injection and be released into the column and ionization source only after the separation and ionization of THMs is completed.

The sampling rate of this method was approximately 20 sample/hr. The MDL values for $CHCl_3$, $CHBrCl_2$, $CHBr_2Cl$ and $CHBr_3$ were 2 ng/L, 4 ng/L, 4 ng/L and 8 ng/L, sequentially. The mean % recoveries for THMs ranged from 3% to 8%. The % RSD for THMs ranged from 1.4% to 4.4%. In the concentration range 25 ng/L to 25 μg/L, the % RSD deviation for each compound was below 9.5% and the r values for THMs were better than 0.993. Over the concentration range 1-20 μg/L, the % RSD ranged from 0.45% to 5.4% and the r values ranged from 0.994 to 0.999. For a 4 μg/L THMs standard solution, the % RSD for each compound ranged from 1.4% to 4.4% and the % difference between the theoretical and experimental concentration ranged from 3.0% to 8.3%. The % RSD each THM species was less than 9.5%.

The major advantages of this method are (1) true on-line real-time monitoring for THMs in drinking water, (2) the separation time was significantly less than the conventional P&T GC method, (3) all the advantages of MS as a detector. Perhaps the largest disadvantage is the expense and the fact that the instrumentation is not commercially available.

SPME GC-MS

Wang and Tan (2000) used a HS-SPME GC-MS method to measure THMs concentrations in water samples from two reverse osmosis/deionization (RO/DI) systems. The MDL estimates for $CHCl_3$, $CHBrCl_2$, $CHBr_2Cl$ and $CHBr_3$ were 0.04 μg/L, 0.04 μg/L, 0.08 μg/L, 0.09 μg/L, respectively. The % RSD ranged from 2.5% to 6.2%. The mean % recoveries ranged from 98% to 100%. The method was done with commercially available instrumentation and the chromatography derived from USEPA methods. The sill level of the analyst is estimated to be medium-high, requiring a knowledge or SPME, chromatography and MS methods. The analysis time is 20 minutes per sample giving an approximate sampling rate of 3 sample/hr.

Stack et al. (2000) used HS-SPME and GC-MS to measure the concentration of THMs in drinking water. An HP 5890 GC was interfaced to an HP 5971A MS. An HP-1 column (50 m × 0.2 mm i.d. × 0.5 μm f.t.) was used with He as the carrier gas (head pressure 29 kPa). The GC-MS was operated in SIM. The MS interface temperature was 280°C with the source temperature at 180°C. The temperature program was held for four minutes at 40°C then increased at 15°C/min to 220°C where it was held for one minute. A 0.75 mm injector liner was used and the GC injection mode was splitless.

The SPME fibers were conditioned at 250°C for one hour prior to use. Samples (2 mL) were transferred to SPME vials which contained 250 μL saturated NaCl solution and a magnetic stir bar. The desorption time was two minutes and the desorption temperature was 220°C. The extraction time was 20 minutes and the chromatographic analysis time was 17 minutes for a total analysis time of 37 minutes.

The MDL estimates for $CHCl_3$, $CHBrCl_2$, $CHBr_2Cl$ and $CHBr_3$ were 2.8 μg/L, 1.4 μg/L, 1.0 μg/L and 1.2 μg/L, correspondingly. The % RSD ranged from 0.9% to 6.2%. The mean % recoveries varied by 20%. All analytes exhibited good linearity. Use of a 100 μm f.t. PDMS fiber reduced analyte loss from the fiber. The method was carried out with commercially available instrumentation and materials. The chromatography was derived from USEPA methods. The skill level of the analyst is medium-high, requiring knowledge of SPME techniques and GC-MS analysis. The analysis time was 30 minutes giving a sample analysis rate of 2 sample/hr.

Guidotti and Ravajoli (1999) proposed a method for measuring the concentrations of THMs using SPME and GC-MS. The SPME fiber was a 100 μm of PDMS. The GC-MS was an HP 5890 GC with HP 5971 MS. The column was an HP-5MS (30 m × 0.25 mm × 0.25 μm f.t.). The carrier gas was He (1.2 mL/min). The injection mode was splitless and the injector temperature was 230°C. The purge valve was set for three minutes. The initial oven temperature was 40°C for five minutes, increased to 90°C at 10°C/min followed by an increase at 15°C/min to 200°C and held for ten minutes. The MS transfer line was 280°C and the dwell time was 50 ms. The following m/z values monitored for each THM: $CHCl_3$ (83 amu, 85 amu), $CHBrCl_2$ (85 amu, 129 amu), $CHBr_2Cl$ (129 amu, 127 amu) and $CHBr_3$ (173 amu, 93 amu). The MDL estimates for $CHCl_3$, $CHBrCl_2$, $CHBr_2Cl$ and $CHBr_3$ were 0.125 μg/L, 0.150 μg/L, 0.090 μg/L, and 0.105 μg/L, respectively. The % RSD ranged from 7.8% to 13.2%. The mean % recoveries varied by 20%. All analytes showed good linearity.

The method was carried out with commercially available materials and instrumentation. The chromatography was derived from USEPA methods. The skill level of the analyst is medium-high, requiring a working knowledge of SPME and GC-MS methods. The SPME exposure time was 15 minutes. The analysis time is 50 minutes per sample. The sample analysis rate is 1.2 sample/hr.

A Simple GC-MS Method for THMs in Drinking Water

Brush and Rice (1994) reported a simple method for measuring THM concentration in drinking water that uses a benchtop GC-MS system. An HP methyl silicone column was used for GC separation. The GC-MS parameters were as follows: The split injection ratio was 10:1. The GC injector temperature was 200°C. The oven temperature was initially set at 30°C for one minute, increased at 10°C/min to 50°C where it was held for 1.2 minutes. The GC-MS interface was 200°C with a 0.75 minute solvent vent delay. The mass scan range was 35 amu to 260 amu. The target ions were at 77 amu, 83 amu, 129 amu, 173 amu in single ion mode. The mass abundance threshold was 300 amu.

Standard mixtures containing each THM and 2-bromo-1-chloropropane as the internal standard were prepared at 100 μg/mL in pentane for detector calibration. A 5000 mg/L standard containing THMs in MeOH was prepared for extraction efficiency studies. The analytical procedure for the GC-MS calibration was as follows. A 100 mg/L standard solution containing four THMs and internal standard was analyzed to determine detector response. Peak areas obtained from reaction mixtures were then used to determine the extraction efficiency of the four THMs and to quantitate the amount of each compound in the reaction mixture. The method was carried out with commercially available instrumentation and materials. The skill required of the analyst is rated at medium-high requiring a knowledge of GC-MS. The time required for analysis is 75 minutes giving a sample analysis rate of 0.8 sample/hr.

Inertial Spray Extraction With Ion Trap-MS for $CHCl_3$

St. Germain et al. (1995) proposed a method with a new interface for VOC analysis directly in aqueous solutions without prior sample treatment. The device consisted of a nozzle directed vertically downward in a spray chamber, which is in turn was coupled through a He jet separator to an ion trap (IT)-MS. The nozzle reduced the liquid sample to an aerosol using He carrier gas and rapid sample injection. The large surface area of the aerosol allowed rapid equilibration of the VOCs between the gas and liquid phases. The aerosol droplets were impinged by their inertia on the floor of the spray chamber, where they coalesced for subsequent removal from the interface. The VOCs were carried from the interface by the He flowing countercurrent to the spray. The combined VOC-He mixture passed through a jet separator, which removed most of the carrier gas, and the concentrated VOCs were then analyzed by an IT-MS. The VOCs in their experiments included the $CHCl_3$. The results showed that the MDL for $CHCl_3$ was 7.5 µg/L.

The inertial spray extraction method was accomplished with a combination of commercially available and non-commercially available instrumentation and materials. The level of expertise required of the analyst is expected to be high, requiring the ability to construct specialized devices and GC-MS. Details relating to the specific analysis of THMs were not given.

Electronic Noses in Water Analysis

Yinon (2003) reviewed the different types of electronic noses that are available. Electronic noses are becoming sensitive to a wider range of analytes, have good selectivity for multicomponent analysis. They have exhibit improved analyte recognition over many detectors. A typical electronic nose is composed of a chemical-sensing system and a pattern-recognition system. This two component system leads to a "chemical fingerprint" when the analyte is detected by the electronic nose. When many chemicals are presented to the sensor, a database of these chemical fingerprints may be developed. One particular application of these electronic noses is as detectors for various types of GC, MS and ion mobility MS applications.

There have been several types of electronic noses developed. One type of the electronic nose uses fluorescent polymers that react with VOCs chemicals and function as a detector. Another type of electronic nose is based on the surface acoustic wave (SAW) detector. The basis of operation of the SAW detectors is the changes in the surface wave characteristics of an indicator sensitive to the properties of the vapor. The SAW-based GC detector has been applied to detecting explosives using a resonator crystal operating at 500 MHz. A frequency shift occurs that is proportional to the mass of the analyte, the temperature of the crystal, and the chemical nature of the crystal surface when condensable analyte vapors impinge on the active area of the SAW crystal. Analytes that are able to condense on the crystal at a given crystal temperature will be detected. This is because the specificity of the SAW detector is based on the temperature of the crystal surface and the vapor-pressure characteristics of the analyte. This SAW detectors have been used to analyze dinitrotoluene (DNT). The system was able to complete an analysis in 10 to 15 seconds using a DB-5 column (1 m × 0.18 mm), thus a fast GC analysis. The SAW detector was coated with the carbosilane polymers. The MDL for DNT was estimated at 92 ng/L.

Stuetz (2001) has applied electronic nose based sensor arrays to the analysis of water and wastewater. Experimental work was done using two laboratory-based systems—the eNose model D and eNose 4000, (Marconi Applied Technologies, Chelmsford, U.K.). Each of these eNose devices consisted of 12 conducting polymer sensors. The eNose instruments were calibrated against a reference standard of 1,2-propanediol, with filtered N_2 used to purge the system during calibration and sample analysis. Samples were purged with N_2 in a temperature-controlled flow cell (25°C); the head space gas was transferred through a temperature-controlled HS transfer line (30°C) to the sensor array module (35°C) for analysis (eNose 5000, Marconi Applied Technologies).

Experimental results showed that an external flow cell could be used to generate a dynamic head space for subsequent sensor array analysis once a reproducible head space has been established. Differences between tainted and untainted water and between different wastewater types using an online flow-cell can be obtained by the dynamic head space sampling coupled to the on-line sensor array. It is possible that such a technique could be applied to measuring the concentrations of THMs and HAAs in drinking water, but further work needs to be done. The analysis time is in seconds, and this method is also very sensitive and not very expensive.

Staples (2000) developed a new version of an electronic nose, the zNose fast GC instrument, for food and beverage analysis. The zNose has many advantages over a previously used eNose sensor. With the zNose fast GC instrument, it is possible to perform on-line quantitative measure of quality for food, beverages, cosmetics and other products, such as drinking water. The zNose system is also based on a SAW detector, providing recognizable images of vapor mixtures containing hundreds of different chemical species. It is possible to perform analysis in ten seconds over a wide range of concentrations. The zNose fast GC system has exhibited picogram sensitivity in some applications and is relatively easy to use and calibrate. Two types of this instrument are commercially available handheld or bench top instruments.

Analysis using the zNose system is a two-step process. In a first step, vapors exiting inlets are concentrated in a Tenax trap. In step two, the trap is heated and released vapors are concentrated on the head of the relatively low temperature (40°C) capillary column. When the column temperature rises linearly to its maximum temperature, different chemical species are released from the sample. They travel through the column and are collected on the surface of a temperature controlled SAW crystal. The SAW crystal consists of an uncoated 500 MHz acoustic interferometer or resonator bonded to a Peltier thermoelectric heat pump with the ability to heat or cool the quartz crystal. Only electrical power is necessary for zNose to work because it carries its own supply of He in a tank capable of supplying enough gas to capture more than 300 VaporPrint images. The instrument includes Windows software to supply macro instructions to the internal programmable gate array microprocessor.

Colorimetric Methods

Most of the colorimetric methods reviewed in this report are based on the Fujiwara reaction. The original Fujiwara reaction has been modified several times. Colorimetric methods based on this chemistry can be used for analyzing both THMs and HAAs concentrations in water. Reckhow and Pierce (1992) described the Fujiwara chemistry in detail. The mechanism is shown in Figure 2.8. Nitrogenated polyenes are formed when dihalogenated or trihalogenated species react with pyridine. N-alkylated pyridines were formed when a polyhalogenated compound undergoes nucleophilic substitution by pyridine. This reaction is first order in pyridine and alkyl halide. N-alkylated pyridines then undergo alkaline ring cleavage and a hydroxylated intermediate is formed by attack at the pyridinium-2 position. Hydrolysis of Hythis compound yields the salt of the amidine and an absorption maximum at 420 nm. If the pyridine concentration was high, an alkylpyridinium intermediate is formed. Hydrolysis of this compound yields another sodium salt of amidine and an absorption maximum at 530 nm. Each of the absorbing products will slowly decompose to form glutaconaldehyde with an absorption maximum at 370 nm.

Two types of alkaline pyridine tests were discussed in the report. The first type used two separate phases: pyridine and aqueous hydroxide. In this case, the hydroxide concentration was very high. The second type was the single phase type. Single phase solutions of alkaline pyridine may be obtained by using dilute aqueous hydroxide or phase transfer catalysts such as tetrabutylammonium hydroxide (TBAH). These methods had a limited base capacity. There are

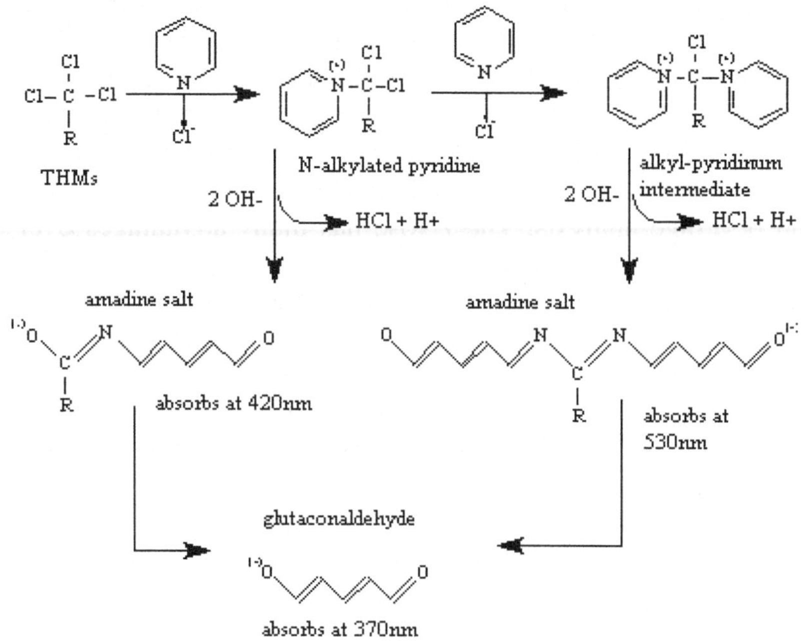

Figure 2.8 The classical pathway for the Fujiwara reaction (Adapted from Reckhow and Pierce 1992)

several different variables that can be changed during the experiment to optimize a particular method: (1) temperature (typically between 70 to 100°C), (2) heating time (typically, three to ten minutes), and (3) the concentration of NaOH or pyridine.

The most sensitive colorimetric methods based on the Fujiwara reaction were for the compounds containing a carbon bonded to three halogen atoms. For instance, MCAA and MBAA cannot be detected by most of the methods because no reaction occurs between monochlorinated compounds and pyridine. The selectivity of the methods based on the Fujiwara reaction is questionable. Usually chlorinated drinking waters contain two major by-products: the THMs and HAAs and four minor by-products: chloral hydrate, trihaloacetones, trichloroacetonitrile, and chloropicrin. Typically only HAAs and THMs will be present in significant concentrations.

Much research has been devoted to detecting TCAA in blood and urine (Ogata, Takatsuka and Tomokuni 1970; Tanaka and Ikeda 1968). Research was extended further for the determination of TCAA in water and air. Some experiments were conducted only for determination of TCAA, and these methods excluded interference with other polyhalogenated compounds. To do this, the Fujiwara reaction was modified or completely new methods were introduced. These experiments were generally very easy to carry out and helped operators distinguish between HAAs and THMs. These colorimetric methods had advantages over other techniques: most notably they are inexpensive to carry out and are relatively simple in terms of procedure.

The majority of methods used pyridine as a reagent in the Fujiwara reaction. Unfortunately, this reaction produces hazardous waste products. Some researchers have used pyridine derivatives in their work that are less toxic and less volatile. Only THMs have been analyzed using

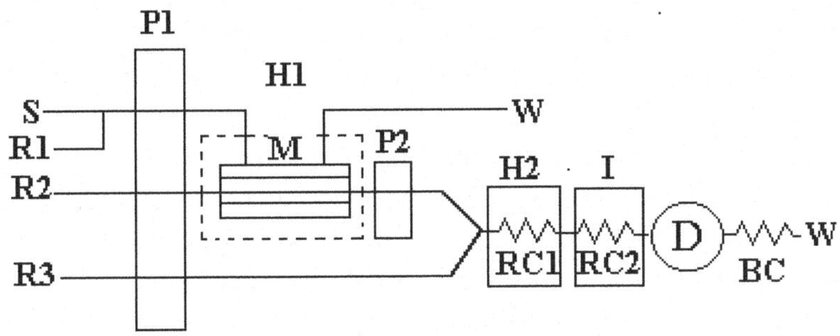

Figure 2.9 Schematic diagram of continuous flow-FIA analyzer with fluorometric detection. S is the sample at 1.48 mL/min; R1 is 10% sodium sulfite at 1.28 mL/min; R2 is 0.2M sodium hydroxide at 0.96 mL/min; R3 is 30% nicotinamide at 0.93 mL/min; P1 and P2 are peristaltic pumps; M is the membrane separation unit at 50°C; RC1 is the reaction coil in water bath at 98°C; RC2 is the cooling coil in ice bath; D is a fluorescence detector with excitation wavelength 372 nm and emission wavelength at 467 nm; BC is the back pressure coil and W is waste (Adapted from Aoki and Kawakami 1989).

pyridine derivatives, no literature was found applying this technique to HAAs analysis. Preliminary research has shown that nicotinamide can be used instead of pyridine to analyze HAAs. Nicotinamide seemed to be a weaker reagent then pyridine as observed by the decreased sensitivity of the methods that employed it. Using fluorescence rather than spectrophotometry may improve sensitivity in nicotinamide based methods. Minor shifts in wavelength were observed when using nicotinamide; the maximum shifted from 368 nm to 352 nm.

Reckhow and Pierce (1992) conducted a detailed study of THM methods based on the Fujiwara reaction. A P&T/one-phase method was proposed giving MDL values in the range from 8 to 12 µg/L with % RSD values ranging from 15% to 22%.

Continuous Flow-Flow Injection Analysis With Fluorometric Detection

Aoki and Kawakami (1989) proposed a continuous flow-FIA method using a membrane separator and fluorescence detector to measure the TTHM concentration in drinking water. The membrane separator used a double-tube separation system with an inner and outer tube made of microporous Teflon (PTFE) membrane. The flow injection manifold and flow rates were as illustrated in Figure 2.9.

The chemical basis of the method was that sulfite ion (R1) decomposed the free available chlorine and chloramines. The THMs permeated the microporous PTFE and were transported by NaOH (R2) to react with nicotinamide (R3). A fluorescence detector was used to detect the products. The time required to reach maximum intensity was eight minutes.

Several factors were investigated to optimize the method. For example, the length of the tubular membrane played an important role on the sensitivity of the method. Maximum fluorescence intensity was obtained using a 40 cm membrane. Once the length of the tube was fixed, the effect of temperature on the membrane separator was studied. The optimum fluorescence intensity was obtained at 50°C. Increasing the temperature of the membrane separator seemed to increase signal stability, likely via degassing of dissolved air in the reagents.

Fluorescence intensity increased with an increase in the nicotinamide concentration. However, the solubility of nicotinamide in water at room temperature (25°C) is such that it cannot be prepared at greater than 40%. Ultimately, to compensate for minor changes during the year, a nicotinamide concentration of 30% was chosen. Studies were conducted over the concentration range from 0.1 M to 0.4 M NaOH to find the optimum NaOH concentration. Maximum intensity was obtained at 0.2 M NaOH. The temperature of the reaction coil one (RC1) was studied. Increasing the temperature above 60°C dramatically increased the fluorescence intensity. Above 98°C, air bubbles began to form in the flow cell, decreasing stability. Finally, the optimum temperature of the (RC1) was 98°C.

The method for $CHCl_3$ exhibited a wide linear range (r>0.999) from 10 µg/L to 10 mg/L. For $CHCl_3$, the MDL detection limit was 0.8 µg/L and the % RSD was 4.4% at 20 µg/L. The relative sensitivities of $CHCl_3$, $CHBrCl_2$, $CHBr_2Cl$ were 100%, 65% and 46%, in that order.

This method was used to determine the TTHM in chlorinated waters at the Niwakubo water treatment plant. The results for measuring TTHMs based upon $CHCl_3$ addition were compared to a LLE GC-ECD method. At the 20 µg/L and 40 µg/L concentration levels, the two methods gave comparable results. The analysis time for the FIA system was 15 minutes per sample.

McKenzie, Studabaker and Carter (1997) developed a rapid colorimetric method for determination of total THMs in drinking water. This method is based on the Fujiwara reaction, in which an organohalogen, pyridine, and hydroxide ion react to give a colored solution. The method provides results of 12 samples within 15 minutes of sample collection. 3 M Empore active carbon SPE filters were used. Attaching the Empore filter directly to a tap and filtering the sample using line pressure can collect samples. THMs are then eluted from the filter with pyridine, using a manual dispenser. NaOH was added to the eluent and heated in the boiling water for two minutes and then cooled at room temperature. A pink color indicated the presence of total THMs and the intensity of the color was proportional to the concentration of THMs. Sample concentrations were determined by reference to a colored chart calibrator or a standard curve. The absorbance was measured in a spectrophotometer at 531 nm. The MDL for total THMs was 5 µg/L. The mean % recovery ranged from 74% to 100%. The % RSD values were ~10% for samples in the concentration range 5 µg/L and 80 µg/L. This method is inexpensive with a short analysis time. Operator skills are not particularly high. It can be readily automated using FIA or SIA techniques.

Hach Company (Lord 1999) has a commercially available colorimetric kit for measuring HAAs and THMs in drinking water. This method does not produce hazardous waste products compared to the method based on the Fujiwara reaction. Instead of pyridine, researchers used N,N-diethylnicotinamide and G-amido acid. The exact concentrations and quantities were not disclosed in this paper. N, N-diethylnicotinamide was mixed with polyhalogenated compound in the presence of alkali. The solution was heated for five minutes and cooled. G-amido acid was added to the solution and let stand for 15 minutes to form an orange dye. A spectrophotometer was used to record absorbance at 515 nm. The MDL for this method for TTHMs and total HAAs was 6 µg/L. For HAAs, this would not include monohalogenated species. There were no details on accuracy and precision. The analysis time for this method is 30 minutes. The method does not require high operator skills and is inexpensive. Both THMs and HAAs can be analyzed by this colorimetric method. This method can be readily automated by FIA.

SUMMARY OF THM METHODS

There are many papers in the literature pertaining to analysis of THMs. The majority of the THMs analysis methods are GC based. The GC approach works well for THMs allowing the use of well established ECD and MS. In addition the GC approach allows other detectors to be explored such as the PDPID and zNose sensor arrays. There are also several colorimetric methods that can be readily automated using FIA, SIA and similar technologies that can enhance the GC methods. A compilation of the MDL, accuracy and precision data as well as other characteristics is presented in Table 4.1 at the end of the Report.

Table 2.1
Reported MDL, accuracy and precision results for THMs using USEPA Methods

Method	Chemical species	MDL (µg/L)	Mean % recovery	% RSD
USEPA 502.2	$CHCl_3$	0.02	98	2.5
P&T-	$CHBrCl_2$	0.02	97	2.9
GC-PID-Hall	$CHBr_2Cl$	0.8	97	2.8
	$CHBr_3$	1.6	106	5.2
USEPA 524.2	$CHCl_3$	0.03	90	6.1
P&T-	$CHBrCl_2$	0.08	95	6.1
GC-MS	$CHBr_2Cl$	0.05	92	7.0
	$CHBr_3$	0.12	101	6.3
USEPA 551	$CHCl_3$	0.002	80	2.4
LLE	$CHBrCl_2$	0.006	99	2.7
GC-ECD	$CHBr_2Cl$	0.012	104	1.5
	$CHBr_3$	0.012	109	0.7
USEPA 551.1	$CHCl_3$	0.080	99	2.90
LLE	$CHBrCl_2$	0.068	89	4.07
GC-ECD	$CHBr_2Cl$	0.068	85	3.38
	$CHBr_3$	0.020	96	2.76

Table 2.2
Types of commercially available SPME fibers (Adapted from Sigma Aldrich 2003)

SPME fiber	Type	Polarity
Bare fused silica	Adsorbent	Unknown
7 μm PDMS	Absorbent	Non-polar
30 μm PDMS	Absorbent	Non-polar
100 μm PDMS	Absorbent	Non-polar
85 μm PA 65 μm	Absorbent	Polar
PDMS-DVB, StableFlex	Adsorbent	Bipolar
65 μm CW-DVB, StableFlex	Adsorbent	Polar
55μm /30μm DVB/CAR-PDMS, StableFlex	Adsorbent	Bipolar
85 μm CAR-PDMS, StableFlex	Adsorbent	Bipolar

Table 2.3
Comparison of the MDL, accuracy and precision studies for the SCMS-GC-ECD and the SCMS-GC-PDECD methods for measuring THM concentrations in water

Parameter	SCMS-GC-ECD	SCMS-GC-PD(ECD)
MDL Total THMs (μg/L)	6.9	2.6
% Error on Slope (Sensitivity)	5	5
Mean % Recovery	120	105
% RSD	4	4
r	0.995	0.995

CHAPTER 3
OVERVIEW OF EXISTING ANALYTICAL METHODS FOR MONITORING HAAs IN DRINKING WATER DISTRIBUTION SYSTEMS

ENVIRONMENTAL PROTECTION AGENCY METHODS

There are four USEPA Methods commonly applied to measure HAA concentrations in drinking water. USEPA Method 552 (1990b) and 552.2 (1995d) use LLE and derivatization of the HAAs to form their corresponding methyl esters. The methyl esters of the HAAs are then analyzed via GC-ECD. USEPA Method 552.1 (1992) employs ion-exchange liquid solid extraction, derivatization to form the corresponding methyl esters, followed by analysis using GC-ECD. USEPA 552.3 (2003b) uses micro-LLE (with methyl tert-butyl ether (MTBE) or tert-amyl methyl ether (TAME)), derivatization and GC-ECD for analysis to measure the concentrations of nine HAAs. Table 3.1 shows the expected MDL, accuracy and precision estimates for these four USEPA Methods. These methods are well established for measuring HAA concentrations in drinking water. However, the derivatization steps that make GC analysis feasible are very cumbersome and not easily adapted to on-line monitoring.

ALTERNATIVE METHODS FOR HAAs

There are several methods appearing in the literature for measuring HAA concentrations in drinking water. The methods have been grouped according to the following categories: GC based methods, GC-MS based methods, SPME based methods, HPLC/IC based methods, LC-MS based methods, CE based methods, colorimetric methods and other approaches.

GC Based Methods for HAAs

Nikolaou et al. (2002a) examined a conventional GC-ECD method to determine HAAs in water by acidic MeOH esterification as the derivatization agent instead of diazomethane. The column was a DB-1 (30 m × 0.32 mm i.d. × 0.25 μm f.t.) with splitless injection mode. The carrier gas was He (1.6 mL/min) and the make-up gas was N_2 (46 mL/min) at a column oven temperature 35-220°C. The injector and detector temperatures were 175°C and 300°C, respectively. The MDL values for the HAA species were as follows: MCAA (0.20 μg/L), MBAA (0.05 μg/L), DCAA (0.02 μg/L), TCAA (0.01 μg/L), DBAA (0.02 μg/L). The mean % recoveries for MCAA, MBAA, DCAA, TCAA and DBAA averaged 125%, 101%, 98%, 105% and 120%, respectively. The % RSD for MCAA, MBAA, DCAA, TCAA and DBAA averaged 1.0%, 5.2%, 6.6%, 5.0% and 5.8%, sequentially. The method was carried out with commercially available instrumentation and materials. The method was based on USEPA methods. The skill level of the analyst is medium, requiring a knowledge of basic gas chromatography and derivatization techniques. The sample extraction and preparation time was about 90 minutes plus a 47 minute chromatographic analysis. Thus the total analysis time is about 140 minutes. This gives a sample analysis rate of 0.4 sample/hr.

Humbert and Fernandez (1976) developed a method for determination of TCAA using GC that is based upon making the 3-methyl-1-p-tolyltriazene derivatives. Solutions of TCAA in the concentration range 500 μg/L to 6 mg/L were prepared by diluting a 20% stock solution with water. A PE model F11 GC with ^{63}Ni ECD and a PE chart recorder were used. The packed column was a 2.0 m × 3.2 mm glass tube containing 5% OV-17 on 100/120 mesh Gaschrom O. The injector was set at 180°C. The oven temperature was held isothermally at 80°C. The ECD temperature was set at 200°C. The carrier gas was N_2 at 50 mL/min. This method was developed to measure TCAA concentrations in urine. The TCAA in samples was extracted with diethyl ether and methylated using 3-methyl-1-p-tolyltriazene. The internal standard was 1,3-dibromopropane. The injection volume for the GC was 2 μL of the diluted ether extract. Repetitive analysis of some 50 mixtures of same concentration gave a % RSD of 5.3% for TCAA. Concentrations ranged between 0.5 mg/L and 6 mg/L, relatively high for drinking water considerations. The method was carried out with commercially available instrumentation and was not directly derived from USEPA methods. The method showed good linearity over this concentration range. The MDL and mean % recovery for TCAA was not evaluated. The preparation time was 90 minutes and the chromatographic analysis time was eight minutes. This gives a sample analysis rate of 0.6 sample/hr. Additionally, it should be noted that the technology used for this method is somewhat dated as separation is accomplished with a packed GC column.

Vesterberg, Gorczak and Krasts (1975) developed a method for determination of trichloroethanol and TCAA in blood and urine using GC. The method depended on extraction with isooctane. They used a PE model F11 GC equipped with a ^{63}Ni ECD. The column was stainless steel with inner diameter of 1.9 mm, 210 cm long, and was packed with 15% SE 30 on Chromosorb W, A.W. 100 m to 120 m. For $CHCl_3$ analysis, the inlet, column, and detector temperatures were 150°C, 80°C, and 200°C, respectively. For the analysis of TCAA, the inlet, column and detector temperatures were 175°C, 125°C, and 200°C, correspondingly. The carrier gas was N_2 at 30 mL/min. For TCAA, the MDL was estimated at 2 mg/L. The retention time was 1.5 minutes for both compounds. A simple HS procedure was developed after esterification of TCAA to increase sensitivity. While the method was done with commercially available instrumentation and materials, the total preparation/analysis time was much too long at approximately 26 hours. In addition, the technology is somewhat dated as with the previous method.

Donnell et al. (1995) developed a GC based procedure for the determination of TCAA in urine. The GC was an HP 5890 with an ECD and an HP 7673A automatic liquid sampler. The GC columns were a DB-17 (30 m × 0.32 mm i.d. × 0.25 μm f.t.) and DB-5.625 (30 m × 0.32 mm i.d. × 0.25 μm f.t.). The injection mode was split injection. The injector temperature was 240°C and detector temperature was 300°C. The temperature program was initially 55°C for 0.5 minutes before being increased at 3°C/min to 80°C, followed by increasing the temperature at a rate of 50°C/min to 280°C, where it was held for four minutes. The carrier gas was H_2 at 2 mL/min. The split vent was equal to 40 mL/min and detector make up gas was 5% methane in Ar at a rate of 80 mL/min.

This method was based on the derivatization of TCAA to its methyl-ester using MeOH/borontrifluoride reagent and heating for 30 minutes at 60°C. After cooling, the sample was extracted with toluene for one hour on a rotator and analyzed by GC. The results were compared with those obtained using a Fujiwara-based colorimetric method. The concentration range was from 0.1 to 100 mg/L. The recovery of TCAA-methyl ester in water was low at 48.8% but reproducible with a standard deviation of 1.9%. In urine, the absolute recovery was 48.5% with a standard deviation of 2.5%. The mean % recovery in urine was 99.6%. The % RSD ranged from 2.2% to 4.9%. The procedure showed good linearity over the range 0.4 to 100 mg/L. The MDL estimates for the method were relatively high at 0.05 mg/L. The method was carried out with commercially available instrumentation and materials. The skill level of the analyst is estimated as medium. The analysis time is 130 minutes giving a sample analysis rate of 0.4 sample/hr.

Xie, Reckhow and Springborg (1998) developed an acidic MeOH methylation procedure for measuring HAAs and ketoacids using acid MeOH derivatization. They compared this method to the diazomethane method. For HAAs the GC conditions were as follows: The column was a DB-1701 (30 m × 0.32 mm × 0.25 μm f.t.). N_2 was the carrier gas at 1 mL/min. The injection mode was splitless with a 0.5 minutes purge time. The injector temperature was 160°C. The detector was a ^{63}Ni ECD held at 300°C. The oven temperature was 37°C for 21 minutes then increased at 11°C/min to 136°C and held for three minutes, increased to 236°C at 20°C/min and held for three minutes. The total run time was 40 minutes.

The procedure for HAAs involved dechlorinating a 30 mL water sample with sodium sulfite and adding sodium sulfate to enhance extraction efficiency of the acids from the aqueous phase to the organic solvent. Each sample was then extracted with 4 mL of MTBE, spiked with 300 ng/L of dibromopropane as the internal standard and agitated for 15 minutes with a mechanical shaker. For the acidic MeOH derivatization, 1 mL of the extract was transferred to a 20 mL vial containing 2 mL of MeOH and 0.1 mL of concentrated sulfuric acid and kept in a 50°C water bath for one hour. After addition of 5 mL of 10% sodium sulfate solution and 1 mL of fresh MTBE, an aliquot was analyzed by GC. For the diazomethane derivatization, 1 mL of the organic extract was dosed with 0.25 mL of diazomethane solution. After the residual diazomethane was removed, the extract was analyzed by GC.

Their results showed that the gas chromatogram of the acidic MeOH-methylated sample gave a cleaner baseline and fewer other peaks. For samples with concentrations of 50 μg/L or higher, TCAA data obtained with the diazomethane method were not reliable because of the presence of chromatographic interferences. The researchers inferred that the acidic methylation method might be less susceptible to certain types of chromatographic interferences. The MDL values for the HAAs were: DCAA (0.2 μg/L), TCAA (0.09 μg/L), MBAA (0.4 μg/L), DBAA (0.07 μg/L). The mean % recoveries were as follows: DCAA (111%), TCAA (109%), MBAA (118%), DBAA (108%).

HS-GC Methods for HAAs

Cancho, Ventura and Galceran (1999) monitored the behavior of halogenated DBPs including THMs and HAAs in the water treatment plant of Barcelona. HAAs analysis involved acidic MeOH esterification technique followed by HS-GC. For HAAs, 30 mL of sample, 5 μL of an ether solution of 2-bromopropionic acid 60 mg/L, 3 mL of concentrated sulfuric acid, 12 grams of anhydrous sodium sulfate, 3 g of copper sulfate pentahydrate and 2 mL of MTBE were added in the 40 mL USEPA glass vials. The vials were sealed and shaken for two minutes and allowed to stand for five minutes. A 900 μL aliquot of the MTBE extract was transferred into a 10 mL vial containing 2 mL MeOH/ sulfuric acid (10%), and 1 μL of an ether solution of 2,3-DBPA 110 mg/L was added. The vials were placed into a 50°C water bath for one hour and then cooled to 4°C for ten minutes. Then 5 mL of 10% $CuSO_4$ / 5% Na_2SO_4 was added. The samples were shaken for two minutes. A 400 μL aliquot of MTBE extract was transferred to a 2 mL vial. The internal standard (4 μL of a mixed ether solution of bromochloromethane and 1,2 dibromopropane 10 mg/L) was added. The injection volume for GC was 1 μL of the MTBE extract. The preparation time was 1.5 hours.

The GC analysis of the HAAs was as follows: 1 μL of MTBE extract was injected manually. The splitless mode was set for 60 seconds. The injection temperature was 175°C. The column was a DB-1 (30 m × 0.25 mm × 1.0 μm f.t.). The carrier gas was He at 140 kPa, and the make-up gas was N_2 at 110 kPa. The ^{63}Ni ECD temperature was 300°C. The oven temperature was initially set at 35°C for nine minutes, increased to 40°C at 1°C/min and held for three minutes before being increased to 220°C at 6°C/min and held for ten minutes. The method used commercially available instrumentation and materials. The chromatography was largely derived from USEPA methods. The skill level of the analyst is rated at medium, requiring a working knowledge of GC and derivatization methods. The sample analysis time includes a sample preparation time of 150 minutes including a chromatographic rum time of 60 minutes.

Wu, Gabryelski and Froese (2002) combined liquid-liquid microextraction (LLME) and SPME with GC-ECD for the analysis of HAAs in micro-volume drinking water. The GC was an HP6890 with micro-cell ^{63}Ni ECD, and an 8200/SPME autosampler was used for the LLME and LLME-SPME analysis. The carrier gas was He at a flow rate of 0.8 mL/min. The oven temperature was initially 40°C then increased to 70°C at 10°C/min, held for four minutes, increased to 220°C at 15°C/min and held for three minutes. The injector and detector temperatures were 230°C and 300°C, respectively.

The extraction procedure was the same as in the LLME-GC-ECD method, with further adaptation of the method reported by Sarrion, Santos, and Galceran (1999, 2000) discussed later in this report. After extraction, MTBE was placed in a 2 mL autosampler vial and evaporated to dryness. Sodium sulfate (0.16 grams), MeOH (10 µL), and sulfuric acid (30 µL) were added to the dried residue in the vial. The solution was vortex mixed and HAAs were derivatized at 80°C for 20 minutes. HS extraction was performed using a 100 µm PDMS fiber for ten minutes at room temperature (25°C). 2,3-DCPA and 2,3-DBPA were used as surrogate and internal standards, respectively, for LLME and SPME methods.

The chromatographic separation time for nine HAAs is 40 minutes. TCAA is strongly retained and thus exhibits a low response because of band broadening. MeOH and EtOH were used as esterification reagents. The retention times of ethyl-esters were several minutes longer than those of methyl esters. HS-SPME reduced the number of unidentified peaks, giving a cleaner baseline. The responses of most HAAs increased with increased holding time between derivatization and analysis. This suggested that derivatization was not complete after a ten minutes incubation time. Ultimately, the derivatization temperature and time were optimized at 80°C for 20 minutes. The responses of HAAs decreased with increasing volume of sulfuric acid. This could be due to dilution of HAAs. Similar effect was seen by increasing the volume of MeOH. The addition of salt improved the absorption efficiency onto the fiber for all the compounds by shifting the equilibrium towards the gas phase.

Good precision was obtained at a concentration near 200 µg/L and was estimated to be in the range from 0.5%-13%. This is an important factor in this method, good precision can be maintained down to sub µg/L concentrations even with the inherently greater level of imprecision that is probable when working with very small sample volumes. The MDL values for six HAAs were between 0.6 µg/L and 1.2 µg/L.

The analysis was performed with commercially available materials and analytical instrumentation. The basis of the derivatization and analysis is rooted in USEPA methods. The skill level of the analyst is rated at medium-high, requiring knowledge of SPME, derivatization chemistries and basic GC techniques. The analysis time including sample preparation and chromatographic separation is 140 minutes. This gives and estimated sample analysis rate of 0.4 sample/hr.

Benanou, Acobas and Sztajnbok (1998) developed a method for HAAs that based on simultaneous extraction and derivatization. The GC was a Varian 3400 with ECD. The column was an HP-1. The injector temperature was 250°C. The detector temperature was 300°C. The carrier gas and the make-up gas was N_2. The oven temperature was 40°C for ten minutes, increased to 180°C at 5°C/min, then increased to 280°C at 15°C/min.

For the simultaneous extraction/derivatization, a column fitted with a frit and filled with one gram of anion-exchange resin was conditioned with 5 mL of 1 M HCl. 10 μL of trichloropropane (internal standard) was percolated at an optimal flow rate of 4 mL/min. The end of the column was closed, and 2 mL of the extract was added. The procedure was repeated with three 2 mL aliquots. The esterified acids were extracted with 5 mL of cyclohexane in the presence of 5 μL of TCP (quantification standard) and then analyzed by GC.

TBAA was not methylated by this technique; it was converted to $CHBr_3$ by acid catalysis. The methyl-ester of MCAA was poorly retained on the selected analytical column and gave a weak response to ECD and was therefore not determined by this method. A resin of 150 μm to 300 μm was optimal for this type of analysis. Cyclohexane proved to be the best solvent for the esters present in the sulfuric acid/ methanol mixture. The % RSD values ranged from 10% to 20% for the method. The MDLs were between 0.1 μg/L and 0.2 μg/L for the six HAAs analyzed.

The method was conducted with commercially available instrumentation and materials. It was not based from USEPA methods. The instrument was a combination of commercially available devices and a column that appeared to be prepared in the laboratory. The skill level required for the technique are rated as medium-high. The analysis time is estimated at 60 minutes leading to a sample analysis rate of 1 sample/hr.

Larger Volume Injection GC-ECD

Drechsel et al. (2001) reported a GC method with large-volume injection (LVI) and programmed-temperature vaporization (PTV) for the analysis of TCAA in water samples. The chemical basis of this method based on the decarboxylation of TCAA that occurs at high temperatures leading to $CHCl_3$ formation. The $CHCl_3$ was measured using the GC and related back to TCAA concentration.

An HP 5890 GC equipped with a Gerstel CIS 3 PTV injector and an ECD was used for the GC analysis. The column was a Varian-Chrompack (30 m × 0.32 mm i.d.). The carrier gas was H_2. The oven temperature was held at 30°C for seven minutes, increased at 10°C/min to 160°C, and increased at 20°C/min to 250°C, where it was held for ten minutes. The PTV injector was equipped with a glass liner packed with Tenax TA, which was used to adsorb the TCAA during injection. A speed-programmable autosampler was used to inject varying volumes of sample between 20 μL and 200 μL. Large-volume injections were performed in stopped-flow mode. The injection speed was 10 μL/min. The initial injection temperature was 50°C and split flow was 600 mL/min. After injection, the injector was heated at 12°C/s to 130°C and held for five minutes with the split valve open for removal of interfering volatile compounds. The split valve was then closed and the injector was heated at 12°C/s to 250°C. After five minutes the split valve was opened. A back-flush device was used for direct injection of aqueous samples with the purpose of keeping water from entering the analytical column during injection.

External standardization was used for calibration. Aqueous solutions of TCAA with concentrations between 1.7 μg/L and 34.4 μg/L were analyzed. The chromatographic peak for the $CHCl_3$ (resulting from TCAA) was asymmetrically shaped. This occurred because the decarboxylation was not instantaneous and a broad band of $CHCl_3$ entered the analytical column. This happened with most wall coated open tubular columns. Thus for this technique, it is necessary to use a column with high retentive power for $CHCl_3$ at low temperatures. This focused the compound in the column inlet during sample transfer, thus narrowing the chromatographic band.

The % RSD ranged from 1% to 9% with MDL values at the 0.1 μg/L level. With drinking water samples a clean-up step was necessary. This step involved initial extraction with a polar solvent prior to analysis. This method practically eliminates complex sample-preparation steps with the exception of the required clean-up step. The analysis time was relatively short compared to other techniques for HAAs. The GC running time was 35 minutes and sample preparation only took about ten minutes. The total analysis time was 45 minutes. The instruments used for the analysis are largely commercially available, but has been modified in house. The estimated skill level of the analyst is rated as medium-high.

GC-MS Based Methods for HAAs

Xie (2001) proposed a new method based on GC-MS in SIM mode to determine all nine HAAs in drinking water. The sample preparation procedure was based upon USEPA Method 552.2. The specific procedure was carried out by adding 1.5 mL of concentrated sulfuric acid, 12 grams of sodium sulfate, the surrogate standard 2,3-DBPA, and 30 mL of water samples were extracted with 3 mL of MTBE containing the internal standard (1,2,3-trichloropropane) for two minutes. Next, 2.5 mL of extract was methylated by adding 1.0 mL of 10% H_2SO_4 methanolic solution and maintained at 50°C for two hours. The extract was cleaned with 4 mL of 10% sodium sulfate solution and then injected into the GC-MS.

The GC was an HP 6890 with HP 5973 MS. The injection mode was splitless. The column was an HP-5MS (30 m × 0.25 mm × 0.25 μm f.t.). The carrier gas flow rate was 1 mL/min. The initial oven temperature was 35°C for ten minutes, increased to 75°C at 5°C/min and held for 15 minutes, increased to 100°C at 5°C/min and held for five minutes, increased to 135°C at 5°C/min, held for two minutes then increased to 185°C at 25°C/min where it was held for two minutes. The injector temperature was 200°C and the transfer line was maintained at 280°C. The temperature of the ion source was 250°C. An automatic liquid sampler was used for sample introduction and injection volume was 3.0 μL.

The following m/z values were monitored corresponding to the methyl-ester derivatives of the HAAs: MCAA (59 amu), MBAA (59 amu), DCAA (59 amu), BCAA (59 amu), TCAA (59 amu), DBAA (59 amu), BDCAA (59 amu), DBCAA (59 amu), TBAA (59 amu). The surrogate was monitored at m/z value 165 amu and the internal standard was monitored at m/z value 75 amu. The MDLs for the HAAs were determined to be: MCAA (0.21 μg/L), MBAA (0.71 μg/L), DCAA (0.07 μg/L), BCAA (0.27 μg/L), TCAA (0.15 μg/L), DBAA (0.04 μg/L), BDCAA (0.12 μg/L), DBCAA (0.20 μg/L) and TBAA (0.83 μg/L). The mean % recoveries for the HAAs were determined to be: MCAA (87%), MBAA (73%), DCAA (92%), BCAA (81%), TCAA (137%), DBAA (93%), BDCAA (144%), DBCAA (165%) and TBAA (117%). Detailed precision data was not reported.

The method was carried out with commercially available instrumentation. It was largely derived from USEPA methods. The skill level of the analyst and instrumentation is rated as high, requiring a working knowledge of derivatization chemistry and GC-MS methods. The time required for sample analysis is about 110 minutes. The sample analysis rate is estimated at 0.5 sample/hr. The major advantages of this method are higher sensitivity, fewer interfering peaks, cleaner base lines, and significantly reduced run time compared to the existing GC-ECD method.

Ozawa (1993) used GC-MS with SIM of brominated HAA concentrations by direct reaction forming difluoroanilide derivatives. Five brominated HAAs (not including TBAA) were determined at μg/L concentration levels with good recoveries from natural waters. The GC-MS was a JEOL

JMS-AX505W. The columns used included a DB-17 (15 m × 0.53 mm i.d. × 1.0 μm f.t.) and DB-5 (15 m × 0.53 mm i.d. × 1.5 μm f.t.). The injector temperature was 220°C, the separator temperature was 230°C and the ion source temperature was 260°C. The carrier gas was He at 20 mL/min. The sample injection volume was 2 μL. The oven temperature was initially 80°C for two minutes and then increased by 8°C /min to a final temperature of 200°C.

The internal standard was deuterated naphthalene solution (2 μg/mL). 50 μL was added to 2 mL of the sample solution. Aliquots of these solutions were injected into the GC-MS system. The difluoroanilide derivatization method was used to analyze aqueous solutions of bromated HAAs. The detection mode was SIM at the m/z values of 249 amu, 327 amu and 405 amu. These m/z values corresponded to the difluoroanilide derivatives of MBAA, DBAA and TBAA, respectively. The derivative of MBAA exhibited a base peak at m/z 129 and , the derivatives of DBAA and TBAA gave base peaks of m/z 156 amu. Chlorinated HAA (DCAA, TCAA, DBAA and TBAA) derivatives were monitored using a similar procedure at m/z values 129 amu, 156 amu and 156 amu, correspondingly.

Calibration plots were prepared by normalizing the area of the ion peaks to the area of the molecular ion peak of the internal standard. The method exhibited good linearity for concentrations ranging over several μg/L. The MDLs for MCAA, DCAA, TCAA, MBAA and DBAA were 0.5 μg/L, 0.5 μg/L, 0.5 μg/L, 2 μg/L and 1 μg/L, respectively in 50 mL water sample. The method was rather insensitive for TBAA as it was not detected at concentrations below 100 μg/L.

The method was conducted with commercially available instrumentation and materials. It was not based upon USEPA methodology. The instrumentation and skill level of the analyst are rated as high requiring a knowledge of derivatization chemistry and GC-MS methods. The sample analysis time is approximately 60-70 minutes giving a sample analysis rate estimated at 0.8 sample/hr.

Kristiansen et al. (1996) used solvent extraction and GC-MS to analyze the HAAs formed in laboratory chlorinated seawater as well as chlorinated seawater and drinking water. The GC-MS was a Varian 3300 GC equipped with an autosampler and Finigan MAT 800 IT-MS. The column was a DB-5 (30 m × 0.25 mm i.d. × 0.25 μm f.t.). The carrier gas was He at a linear velocity of 30 cm/s. A splitless injection temperature was 250°C. The oven temperature was 45°C for five minutes, increased to 100°C at 2°C/min, held for five minutes, increased to 250°C at 20°C/min and held for ten minutes. The mass range from 50 amu to 450 amu was scanned with a cycle time of one second. The run time was one hour.

The extraction procedure for HAAs was as follows: To 100 mL water sample, 34 grams of NaCl were added and the pH was adjusted to less than 1 with concentrated sulfuric acid. The samples were extracted twice with 10 mL of MTBE in separation funnels by shaking vigorously for ten minutes. The combined extracts were kept at -20°C, filtered through glass wool to remove water crystals and concentrated to 1 mL with a gentle steam of purified N_2 gas. The HAAs were methylated by addition of diazomethane in MTBE. The extracts were diluted to 20.0 mL and to 100.0 mL with MTBE before GC-ECD analysis. 1,3-dibromobutane(1,3-DBB) was added to the diluted extract as a syringe standard before chromatographic determination. The sample preparation time was 40 minutes.

The extraction efficiencies of the HAAs were determined by spiking water at a concentration level of 10 μg/L for each of the HAAs. The mean % recoveries for DCAA, TCAA and DBAA were 69%, 70% and 84%, respectively. The % RSD values were 9.4%, 9.5% and 21%, respectively. The recovery of TBAA was low (less than 15%). The MDLs were 0.1 μg/L, 0.15 μg/L and 0.2 μg/L for TCAA, DBAA and DCAA, sequentially.

The method was done using commercially available instrumentation and materials. The method was similar to USEPA based methods. The level of instrumentation and skill level of the analyst are rated as high since a working knowledge of derivatization and GC-MS methods are required. The sample analysis time including derivatization and chromatographic separation is estimated to be 100 minutes, giving a sample analysis rate of 0.6 sample/hr.

Martinez et al. (1998a) used SPE coupled with GC-MS for HAAs analysis. Divinylbenzene disks with quaternary amine as strong anionic exchanger were used in the SPE step. The HAAs analyzed included MCAA, MBAA, DCAA, TCAA, BCAA, and DBAA. The internal standard was 1,2,3-trichloropropane. The GC-MS was an HP 5890 with an HP 5971 MS and an HP7673 auto sampler. The GC was equipped with a split/splitless injector. The GC column was BP-5 (60 m × 0.22 mm × 0.25 μm f.t.). The column was inserted directly into the ion source of the mass spectrometer. The mass range was from 10 amu to 650 amu in the full scan mode. The MS was tuned to m/z 69 amu, 219 amu and 502 amu with perfluorobutyl amine. The oven temperature was 35°C for ten minutes, increased to 170°C at 8°C/min, and then increased to 290°C at 25°C/min. The injector and detector temperatures were 150°C and 280°C, respectively. The carrier gas was He (0.70 mL/min).

The extraction process was as follows: 30 mL of sample were adjusted to pH 0.5 with sulfuric acid, before the extraction step to facilitate extraction of the non-dissociated HAAs. MTBE was used as the organic phase. The HAAs were methylated with diazomethane to produce methyl esters. In the SPE process the sorbent used was divinylbenzene with quaternary amine as the functional group. Once activated, 50 mL of sample water, adjusted to pH 5, was passed through the disk at a flow rate of 10 mL/min using a vacuum system. The HAAs were eluted with 10 mL of 3 M sulfuric acid in MeOH. To increase the extraction efficiency five grams of sodium sulfate were added. The sample injection volume was 1 μL.

Quantitative determinations using this extraction process were made. The linearity of the response was studied between 1 μg/L and 40 μg/L depending on the compound under study. The responses were linear and there was a good correlation for every compound. The following MDLs were reported: MCAA (20 μg/L), MBAA (20 μg/L), DCAA (3 μg/L), TCAA (5 μg/L), BCAA (5 μg/L), DBAA (20 μg/L). The % RSD were between 4% for DBAA and 12% for DCAA. The mean % recoveries were generally low and ranged between 18% and 45%. The following m/z values were monitored: MCAA (77 amu, 108 amu), MBAA (93 amu, 152 amu), DCAA (83 amu, 85 amu), TCAA (117 amu, 119 amu), BCAA (129 amu, 127 amu), DBAA (173 amu, 232 amu).

Clemens and Schoeler (1992) proposed a simple GC-MS method for measuring the concentrations of MCAA, DCAA and TCAA in water. They compared their method to a GC-ECD method for these HAA species. In practice, a 970 mL water sample was adjusted to pH 1.5 with concentrated sulfuric acid and extracted in a light-phase rotary perforator for two hours with 100 mL diethylether. The methylation step involved transferring the sample extracts into calibrated test tubes with ground stopcocks and concentrating them at 50°C without vacuum to about 0.9 mL. Then 50 μL of cooled etheric diazomethane were added. The sample was diluted to 1 mL with diethylether and held at room temperature for one hour. The by-products formed did not influence the chromatography.

The GC column was a CO-Sil 8 CB (25 m × 0.32 mm i.d. × 1.2 μm f.t.). The injection mode was splitless with a 2 μL sample injection volume. The injector temperature was 270°C. The carrier gas was He at 2.0 mL/min. The make-up gas was Ar/CH_4 (95%/ 5%). The ECD was a Carlo-Erba HT 25 held at 300°C. The oven temperature was initially at 45°C for six minutes then increased at 5°C/min to 70°C and held for two minutes. The oven temperature was increased at

10°C/min to 120°C and held for three minutes before being increased at 10°C/min to 180°C. The total chromatography time was 37 minutes.

The GC-MS detector was a Finnigan-MAT, ITD 700 with EI. The retention gap was phenyl-sil (5 m × 0.32 mm i.d.). The column was a DB-5 (30 m × 0.25 mm i.d. × 0.25 μm f.t.). The injection mode was on-column with 8 μL sample volume. The carrier gas was He at 2 mL/min. The oven temperature was 35°C for two minutes, increased at 20°C/min to 75°C and held for six minutes before being increased to 200°C at 10°C/min.

The average recovery for the GC-MS method for MCAA, DCAA, and TCAA were 82%, 92% and 79%, in that order. The % RSD on the analysis for MCAA, DCAA and TCAA was 15.2%, 14.5% and 16.4%, respectively. The MDLs for MCAA, DCAA and TCAA were 1.5 μg/L, 0.120 μg/L and 0.050 μg/L. The mean % recoveries ranged from 64% to 83%.

The SPE-GC-MS method is carried out with commercially available instrumentation and materials. The chromatographic analysis was derived from USEPA based methods, but the use of SPE departed from the standard methods for sample extraction. The level of the instrumentation and skill level of the analyst are rated as high. The time for analysis of a typical sample is 90 minutes giving a sample rate of 0.7 sample/hr.

Scott and Alaee (1998) used an in situ anilide derivatization technique for measuring HAAs in water. The acid anilides were not commercially available but were prepared in the laboratory using two different methods. For TCAA and DBAA, 0.02 moles of the parent acid were placed in 20 mL of diethyl ether containing 0.02 moles of pyridine. Then, 0.02 moles of thionyl chloride was slowly added. The remaining acid chloride solution was added dropwise to 20 mL of cooled $CHCl_3$ containing 0.025 moles. The remaining anilides could not be made in sufficient quantities for recrystallization and subsequent standards using the above method. For these compounds, the following method was utilized: 0.01 moles of the acid was dissolved in 100 mL of distilled water, previously extracted with n-hexane, and containing 1 mL of concentrated HCl and three grams sodium chloride. Then 1.5 mL of 2, 4-difluoroanilides, 30 mL ethylacetate and 2.42 grams of DCC in 2 mL EtOH were added. The mixture was stirred for two hours, 10 grams sodium chloride was added and the phases separated. The combined acetate solution was reduced in volume to dryness. The resulting crystals were the anilide of interest. Generally, all anilides obtained were of 97% purity. All the anilides were characterized using GC-MS. The MS spectra for all mono substituted acid anilides had ions at m/z 101 amu, 129 amu and the parent ion. The polysubstituted acid anilides fragmented to produce ions at m/z 101 amu, 128 amu, 156 amu and parent ions at 243 amu.

Water samples were reduced in volume to 50 mL using a rotoevaporator at 55°C. A blank was also analyzed at the same time. For the blank, 50 mL of the purified water was used along with the reagents and solvent. Up to 750 mL of this water was prepared at a time and was also used for the saturated solutions of Na_2SO_4 and $NaHCO_3$. For the recovery experiments known amounts of the parent acids were dissolved in 50 mL of the laboratory purified water.

An HP 5890 GC with an HP 5971 MS was used for this analysis. The source temperature was 187°C. The column was a DB-17 (30 m × 0.25 mm × 0.25 m f.t.) The injector temperature was 200°C. The oven temperature was 50°C for two minutes, increased at a rate of 6°C/min to 250°C and held for five minutes. The MDLs were in the single μg/L concentration range. The chromatographic analysis took approximately 40 minutes. The % RSD values for MCAA, DCAA and TCAA were 10%. The % RSD values for MBAA and DBAA were considerably higher (41% and 15%, respectively). Commercially available instrumentation was used to carry out the measurements. The reagents were prepared in the laboratory and were not commercially available.

The chemical basis of the method was not based on USEPA methodology. The level of the instrumentation and skill of the analyst is rated as high, requiring knowledge of specialized chemistry and GC-MS methods. The sample analysis time is approximately 135 minutes and the sample analysis rate is 0.4 sample/hr.

In another study, Scott et al. (2000) measured HAAs in natural waters from lakes and rainwater in Canada. The samples were reduced in volume by rotoevaporating to 50 mL. Then one gram of NaCl and 0.2 mL of concentrated HCl were added. This was followed by adding 0.2 mL of 1 M solutions of dicyclohexylcarbodiimide and 2, 4-difluoroanilide in ethyl acetate as well as 25 mL of ethyl acetate. The mixture was stirred for one hour, and five grams of sodium chloride was added and dissolved. The GC-MS analysis was done with an HP 5890 GC and an HP 5971 MS. The column was a DB 17 (30 m × 0.25 mm × 0.25 μm f.t.) The carrier gas was He. During the analysis of each sample set, two standard solutions were also analyzed. The intensity of the peaks at m/z 129 amu for the monosubstituted HAAs, m/z 156 amu for the polysubstituted HAAs and m/z 225 amu for TFA were used to determine the concentrations of the acids. This method is similar to the method describe above. It was conducted using commercially available instrumentation and non-commercially available reagents. The analysis time was decreased to 90 minutes, improving the sample rate to approximately 0.7 sample/hr.

SPME Based Methods for HAAs

Sarrion, Santos and Galceran (1999) used SPME coupled to a Varian 3400 CX GC with Saturn 3 IT-MS for the analysis of HAAs. This method was proposed as a possible alternative to LLE for HAAs. After a ten minute derivatization to form corresponding ethyl esters, HS-SPME was used in conjunction with GC-MS. MDLs on the order of ng/L were reported with % RSD less than 10%. The procedure was applied to tap water and compared to USEPA Method 552.2 with good results.

The acidic ethanol (EtOH) esterification HS-SPME procedure was carried out as follows: The SPME fiber was 100 μm f.t. PDMS fiber conditioned for two hours in the GC injector port at 250°C. A mixture of 30 μL of sulfuric acid and 70 μL of an EtOH solution containing the HAAs were placed in a vial and incubated at 60°C for ten minutes to derivatize the HAAs. After cooling, the vial was placed in a thermostatic bath and 100 μm PDMS fiber was exposed to the HS to extract the HAA ethyl-ester. In developing the HS-SPME optimization, the exposure time, the extraction temperature, the ionic strength and the desorption time were studied to obtain maximum sensitivity GC analysis. For the HS-SPME procedure, the equilibrium time between the stationary phase and the acidic EtOH solution at 25°C for all HAAs derivatives was ten minutes. Increasing the ionic strength by adding 0.6 grams of salt improved the absorption capacity of the fiber coating for all the compounds. The desorption time of two minutes was enough to desorb the compounds completely.

The specific conditions for the IT-MS were as follows: EI positive mode with automatic gain control. The filament emission current was set at 79 μA and the electron multiplier voltage was 2100 V with a modulation amplitude of 2.5 V using perfluorotributylamine as the reference. The transfer line was 270°C and the IT was maintained at 220°C. A solvent delay (2.5 minutes) was used with an acquisition time of 25 minutes. The mass range from m/z value 27 amu to 260 amu was scanned at a rate of 0.8 seconds per scan with ionization time of 500 ms. Two selected ions were monitored for each ethyl haloacetate derivative at the following m/z values: MCAA (m/z 77 amu/94 amu), DCAA (83 amu/85 amu), TCAA (117 amu/82 amu), MBAA

(121 amu/138 amu), BCAA (129 amu/109 amu), BDAA (174 amu/120 amu). The linear range was from 0.2 to 300 g/L. The MDLs were in the concentration range 0.10 to 0.20 µg/L for the HAAs studied. The % RSD estimates ranged from 6.3% to 7.9%. Mean % recovery result was not reported. The method was carried out using commercially available instrumentation and materials and was not based directly on USEPA methods. The level of instrumentation and skill level of the analyst are rated as high. Knowledge of SPME techniques, derivatization chemistry and GC-MS is required. The sample analysis time is 90 minutes including preparation and chromatography. The sample analysis rate is estimated to be 0.7 sample/hr.

Additional research (Sarrion, Santos and Galceran 2000) reported a procedure for derivatization of HAAs to their methyl esters with dimethyl sulfate followed by HS sampling using SPME with GC-IT-MS. The method was applied to MCAA, MBAA, DCAA, DBAA, TCAA, TBAA, chlorodibromoacetic acid (CDBAA), and BCAA. 2,3-DBPA and 1,2-dibromopropane were used as the surrogate standard and internal standard. For evaluating the SPME procedure, water standards containing 200 µg/L of each HAA were prepared.

The separations were done using a Varian 3400 CTX GC-ITS-MS with a DB-5 MS capillary column (30 m × 0.25 mm i.d. × 25 µm f.t.) and He as the carrier gas at 34 cm/s. The oven temperature was held at 40°C for one minute, increased to 60°C at 20°C/min, increased to 120°C at 5°C/min and held for three minutes before being finally increased to 280°C at 25°C/min where it was held for ten minutes. The total run time was 34 minutes.

The specific conditions for the IT-MS were as follows: EI positive mode with automatic gain control. The filament emission current was set at 75 µA and the EMV was 1850 V with a modulation amplitude of 2.5 V using perfluorotributylamine as the reference. The transfer line was 270°C and the IT was maintained at 220°C. During the ethylation studies, the GC-ITS-MS was operated in full scan mode with a mass range from m/z value 27 amu to 325 amu at 0.8 s/scan. Two selected ions were monitored for each ethyl haloacetate at the following m/z values: MCAA (m/z 77 amu/94 amu), DCAA (83 amu/85 amu), TCAA (121 amu/138 amu), MBAA (121 amu/138 amu), DBAA (174 amu/120 amu), TBAA (251 amu/172 amu), BCAA (129 amu/109 amu), BDAA (163 amu/161 amu), CDBAA (209 amu/205 amu). During the methylation studies, the GC-ITS-MS was operated in full scan mode with a mass range from m/z value 29 amu to 260 amu at 0.8 s/scan and an ionization time of 100 ms. Once optimized conditions were established, different acquisition segments during each chromatographic run were applied using a narrow mass range and different scan rates. The response was enhanced by using ionization times of 200 or 400 ms, depending on the analyte. For the methyl haloacetates, two selected ions were monitored at the following m/z values: MCAA (m/z 59 amu/108 amu), DCAA (83 amu/85 amu), TCAA (117 amu/119 amu), MBAA (93 amu/95 amu), DBAA (173 amu/171 amu), TBAA (251 amu/253 amu), BCAA (129 amu/127 amu), BDAA (163 amu/161 amu), CDBAA (209 amu/205 amu). Detailed mean % recovery studies were not reported.

In the HS-SPME procedure, 10 mL of water containing 200 µg/L of each HAA was delivered to a 30 mL glass vial. Five grams of anhydrous sodium sulfate and 100 µL of diethylsulfate were added to the vial. The sample was maintained at 55°C for five minutes with the SPME fiber exposed to the HS above the aqueous solution for the optimized extraction time using 1,2-dibromopropane as an internal standard. It was noted that homogeneous polymer coatings like PDMS and PA fibers did not extract HAAs methyl esters. Dual coated DVB-CAR-PDMS and CAR/PDMS fibers did extract all HAAs, including the mono-halogenated species. Of these fibers, CAR/PDMS gave the highest extraction efficiencies for all HAAs. A sampling time of 25 minutes was used for optimization purposes.

The effect of sample temperature on the derivatization/HS-SPME was studied over the range 35 to 65°C. Increasing the temperature enhanced derivatization reaction yield. As temperature increased the affinity of the HAAs for the SPME fiber coating diminished and the relative responses decreased.

The concentration of the derivatization reagent also affected the reaction yield of HAA methyl esters. The addition of 100 μL of DMS ensured the maximum response for MBAA methyl ester. Tetrabutylammonium hydrogen sulfate (4.7 mM) was used as the ion-pairing agent. Adding modifiers like ion-pairing agents, activated the analytes during derivatization and increased esterification yields. This led to improved sensitivity. Sodium sulfate (3.5 M) was used to increase the ionic strength-thus "salting out" the analytes. Increasing the ionic strength of the solutions, generally decreased the solubility of most organic compounds in water. Increasing the ionic strength increased the concentration of HAA-methyl esters in the vapor phase, making them more available to the HS-SPME sampling approach.

The MDL values were evaluated in full-scan mode at optimized conditions. The MDL values ranged from 17 picograms for DBAA and BCAA methyl esters to 156 picograms for TBAA methyl-ester. The % RSD ranged from 9.8% to 13.9%.

HPLC / IC Based Methods for HAAs

HPLC and IC based methods are becoming more popular for HAA analysis. As a general rule, the method that are reported in the literature are carried out with commercially available instrumentation. There are a wide array of commercially available columns, suppressors and other materials that aid researchers in this area. The analysis times are typically shorter than the GC based methods, typically under one hour for many of these techniques. This shortened analysis time is largely due to the elimination of derivatization steps. These methods are not typically derived from USEPA based methods. Compare to GC methods, the perception is that HPLC and IC based methods require higher skilled analysts. However, this perception is gradually changing as HPLC and IC techniques are becoming more common.

Lopez-Avila (2000) described a method based on a microextraction preconcentration technique of nine HAAs followed by IC analysis with membrane suppressed conductivity detection. The separation column was Dionex AS11 (2 mm i.d. × 250 mm) and the concentration column was a Dionex TAC-LP column (4 mm i.d. × 35 mm). The proposed method used commercially available instrumentation. The MDL values for HAAs ranged from 0.045 to 1.1 μg/L. The major disadvantage of this method was that the steps needed to ensure complete separation and to eliminate the interferences complicate the method considerably. The researchers noted that the method was no longer a "simplified method" and will require an experienced analyst. In addition, the method was not fully optimized. The researchers recommended that the method be further developed so that direct analysis of HAAs without preconcentration is possible.

Sarzanini, Bryzzoniti and Mentasti (1999) examined preconcentration and separation of HAAs using LC. The methods studied included reversed-phase ion interaction chromatography with UV detection and suppressed and non-suppressed anion exchange chromatography. Each method was applied to analysis of HAA5 in drinking water. The ion interaction-based method used a TBA-MeOH or CTA-ACN eluents. The HPLC column was a simple, inexpensive C_{18} column. A major advantage of these methods was the removal of chloride ion (Cl^-) as an interferent. The MDL values for the HAA5 species was reported to be in the range 5 to 400 μg/L.

The anion exchange based methods used either suppressed conductivity or UV detection. The optimum method reported for suppressed conductivity detection used a carbonate based eluent and Ion Pac AS9 column. Cl^- interference was minimized in this method by pretreatment with silver cartridge. MDL values were in the range from 25 to 207 μg/L. A method using anion exchange chromatography with UV detection optimized for an Ion Pac AS10 column was developed. Two eluent systems were proposed for HAAs. Preconcentration techniques were required to reduce MDL values and interferences. Reversed-phase and activated carbon were used as preconcentration agents. Careful control of sample pH during preconcentration allows for minimization of Cl^- and NO_3^- interference. With reversed-phase preconcentration and UV detection, the MDL ranged from 7 to 40 μg/L. With activated carbon preconcentration and suppressed conductivity detection, the MDL values ranged from 16 to 150 μg/L.

Each of these LC based methods used commercially available instrumentation and consumables. The level of instrumentation and analyst skill was rated as medium, requiring a fundamental understanding of LC and IC based analysis. In each case the MDL values were higher than needed for routine on-line monitoring. The analysis time including preconcentration and chromatographic separation was 50 minutes for each method. The sample analysis rate is estimated at 1.2 samples/hr. MCAA was not detected in either method.

Nair, Saari-Nordhaus and Anderson (1994) reported two IC methods based on anion exchange separation with suppressed electrical conductivity detection and anion exclusion separation with UV detection (at 210 nm) for the determination of HAAs and compared both methods for HAA5. They used an Alltech IC system with an Alltech universal anion 300 column (150 mm × 4.6 mm), which consisted of an HPLC pump (Model 325), suppressor module (Model 335) and a conductivity detector (Model 320). For anion exclusion chromatography, the same HPLC pump and injector with a column heater (Model 330) and a UV-VIS detector (Model 204) along with a Wescan Anion Exclusion Column (300 mm × 7.8 mm) were used. The temperature of the column was 35°C. The injection volume for the method was 100 μL. The following MDLs were reported for the ion exchange method: MCAA (8 μg/L), DCAA (16.0 μg/L), TCAA (80.0 μg/L), MBAA (21.0 μg/L), DBAA (30.0 μg/L). The following MDLs were reported for the ion exclusion method: MCAA (70 μg/L), DCAA (8.0 μg/L), TCAA (5.1 μg/L), MBAA (85.0 μg/L), DBAA (90.0 μg/L).

In the suppressed conductivity based method, sodium carbonate (2.2 mM) and sodium hydrogen carbonate (2.8 mM) with flow-rate of 2 mL/min was used as eluent. In the ion exclusion method, the eluent was 1% phosphoric acid with a flow-rate of 0.8 mL/min. Either IC method could be used for the monitoring of HAAs in water, though the MDLs were not as low as needed. Both methods have some co-elution problems that could possibly be overcome. MDLs could further be improved by pre-concentration of the sample.

The methods proposed by Nair, Saari-Nordhaus and Anderson (1994) were done with commercially available instrumentation and consumables. They were not derived from USEPA approved methods. The level of instrumentation and skill level of the analyst were rated as medium, requiring basic knowledge of LC techniques. The analysis time ranged from 40 to 50 minutes, depending on the specific method, and sample analysis rates ranged from 1.2 to 1.5 sample/hr.

Carrero and Rusling (1999) report a technique for HAAs analysis. HPLC was used with an electrochemical detector electrode of pyrolytic graphite coated with a film of the ionomer Nafion and the water insoluble surfactant dididecyldimethyammonium bromide. This HPLC method was used to separate and detect HAA6. MDL estimates were poorer than USEPA Methods 552 and

552.2 but one advantage was that it could be used to measure TBAA concentrations. The major disadvantage is that the electrode is not commercially available and relatively complex to prepare.

Husain et al. (1992) used HPLC for TCAA analysis. The column was a C_{18} column (ChromSphere: 250 mm × 4.6 mm i.d × 10 μm f.t.). A 20 μL sample loop was used on a six-port valve. Aqueous ammonium sulfate (0.15 M) was used as eluent with a flow rate of 1 mL/min at ambient temperature. The absorbance was monitored at 210 nm using a variable wavelength UV-VIS detector. The MDL for TCAA was 10 μg/L. Detailed accuracy and precision results were not reported. Commercially available instrumentation was used for the analysis. The technique was not based on USEPA methods. The analysis time for a sample was approximately 40 minutes, giving a sample analysis rate of 1.5 sample/hr. The level of instrumentation and skill level of analyst is rated at medium, requiring a working knowledge of LC techniques.

Akhtar, Too and Wallace (1997) studied conducting polymer based microelectrodes as HAA sensors. A method for HAAs using a polypyrrole/sulfobenzoic acid (Ppy/SBA) coated electrodes was proposed. These sensors were compared to suppressed conductivity and UV detection. The basis of detection was that the presence of ions facilitates the polymer oxidation/reduction at appropriate potentials. The working electrode was a platinum disc electrode (either 10 μm or 50 μm diameter) and the reference electrode was Ag/AgCl (3 M NaCl) with a salt bridge. The column and suppressor were Dionex Ion Pac AS11 analytical column (4 mm × 250 mm) and AMMS11 membrane suppressor. The eluent was NaOH (21.0 mM) and the regenerant was 12.5 mM H_2SO_4 (flow rate of 2.5-3.0 mL/min.). The injection volume was 50 μL. While the most of the instrumentation used for the method are commercially available, the electrodes are not. The technique is not based on USEPA methods. The level of the instrumentation and analyst skill is estimated to be high. Detailed accuracy and precision studies were not reported. The MDL values measured for the Ppy/SBA microelectrode were as follows: MCAA (1.00 μg/L), DCAA (10.0 μg/L), TCAA (100.0 μg/L), DBAA (100.0 μg/L). The MDLs for the HAAs using the conductivity detection were as follows: MCAA (8.0 μg/L), DCAA (16.0 μg/L), TCAA (80.0 μg/L), DBAA (30.0 μg/L). The MDLs for the UV detection method for MCAA, DCAA, TCAA, and DBAA were 70.0 μg/L, 8.0 μg/L, 5.1 μg/L, and 90.0 μg/L, respectively. In each case, the reported MDL values are too high for practical use.

Vichot and Furton (1994) explored the variables influencing the direct determination of HAAs using reversed phase ion pairing HPLC. They used an HPLC system designed for portability consisting of a Scientific System Inc. (SSI) Model 300 LC pump, a SSI Model LP-21 LO-PULSE pulse controller, an Omega Engineering Type PSW Model 133 8835 gauge, and an E-Lab OMS TECH gradient controller. A Valco injection valve with a 20 μL sample loop was used. Detection was performed with an ISCO V4 UV-VIS absorption detector. The spectrum of the mobile phase was obtained with a Shimadzu UV-2101 PC UV-VIS scanning spectrophotometer. The mobile phase exhibited absorbance maxima at 269 nm, 263 nm and 258 nm. The UV absorbing component of the mobile phase was in equilibrium with the stationary phase of the column until the injected molecules disturbed the equilibrium giving rise to a detector response. The fluorescence of the mobile phase was evaluated with a Perkin Elmer luminescence spectrometer Model LS 50. The columns compared were SGE glass-lined stainless steel (4 mm × 25 cm i.d. × 5 μm particle size) with the following chemistry, pore sizes, surface areas and endcapping: Spherisorb C-18 (80 A, 220 m^2/g, fully); Phenyl (80 A, 220 m^2/g, partly), C-8 (80 A, 220 m^2/g, fully); Hypersil C-18 (120 A, 170 m^2/g, fully); Nucleosil C-18 (300 A, 190 m^2/g, fully), Nucleosil C-4 (300 A, 190 m^2/g, fully).

The effects of temperature, phosphate concentration, hexane sulfonate concentration, acetonitrile concentration, mobile phase pH, and ion interaction reagent concentration were examined. A mobile phase containing 8.22 mM benzyltributylammonium, 0.13 mM hexanesulfonate, 5.66 mM monobasic potassium phosphate, pH 5.2 and 11% acetonitrile gave the best conditions for separating the six acids from acetic to DBAA in less than 20 minutes. TCAA acid was not included in the optimization experiments because it had very long retention time and poor detectability. The flow rate was 0.6 mL/min at a column temperature of 38°C. HAAs were detected at 262 nm. A Spherisorb C-18 column gave the best results for separation of HAAs. The MDL values were in the concentration range 2 to 15 mg/L—much higher than is practical for drinking water monitoring. Details of accuracy and precision were not reported.

Liu et al. (2003) proposed a novel method for the simultaneous determination of nine HAAs in drinking water by IC with large volume direct injection. They used a Dionex 500 IC with GP40 gradient pump, ED40 electrochemistry detector and PeakNet 5.11 chromatography workstation. A Dionex Ionpac AG9HC anion guard column (50 mm × 4 mm i.d.) with Ionpac AS9HC anion analytical (separation) column (250 mm × 4 mm i.d.) and an ASRS-ULTRA anion suppressor (4 mm) were used. The eluent was 28 mM sodium carbonate. The nine HAAs were separated by using a eluent flow-rate gradient which was 0 to 15 minutes at 1.0 mL/min; from 15.1 minutes, increased to 1.2 mL/min to 35 minutes; from 35.1 minutes flow rate returned to 1.0 mL/min. The injection volume was 500 μL. The column temperature was ambient temperature. The total run time was 35 minutes. The following MDLs were determined: MCAA (1.91 μg/L), MBAA (3.11 μg/L), DCAA (4.62 μg/L), BCAA (3.49 μg/L), DBAA (12.50 μg/L), TCAA (5.11 μg/L), BDCAA (28.01 μg/L), DBCAA (38.14 μg/L), TCAA (48.98 μg/L). Good linearity was observed over the 10 to 6000 μg/L concentration levels. The mean % recovery for the HAAs were: MCAA (103%), MBAA (92.2%), DCAA (93.5%), BCAA (106.7%), DBAA (92.5%), TCAA (91.3%), BDCAA (91.5%) DBCAA (95.5%), TBAA (94.7%). The % RSD for the HAAs were: MCAA (4.45%), MBAA (5.66%), DCAA (3.11%), BCAA (4.29%), DBAA (3.32%), TCAA (3.99%), BDCAA (2.86%) DBCAA (2.44%), TBAA (2.15%).

In another study, Liu and Mou (2003) developed a new method for simultaneous determination of nine HAAs, perchlorate ion and some common anions in one run by IC. Two high capacity anion exchange columns were used. One was a carbonate selective column in which nine HAAs as well as fluoride, chloride, nitrite, nitrate, phosphate and sulfate could be well separated, and the other was a hydroxide-selective hydrophilic column in which perchlorate ion in addition to all the above analytes could be determined. The IC was Dionex DX-500 equipped with a GP40 gradient pump, IonPac AS16 analytical column (250 mm × 4 mm) and IonPac AS9HC (250 mm × 4 mm). Two guard columns AG16 and AG9HC were also used. A 500 μL sample loop was used for the injection. The eluents for AS9HC were 28 mM Na_2CO_3 and 11.5 mM Na_2CO_3 and for AS16 column were 100 mM NaOH, deionized water and 10% aqueous MeOH solution. The eluent flow rate was 1 mL/min. The analysis time was 34 minutes. The MDLs for HAA5 were MCAA (1.46 μg/L), MBAA (2.49 μg/L), DCAA (2.16 μg/L), DBAA (2.16 μg/L), TCAA (1.85 μg/L). The % RSD was: MCAA (3.4%), MBAA (8.1%), DCAA (5.2%), DBAA (3.16%), TCAA (2.9%). The mean % recoveries were MCAA (98.8%), MBAA (99.1%), DCAA (98.5%), DBAA (98.0%), TCAA (103.2%).

Each of the methods proposed by Liu and coworkers are done with commercially available instrumentation and consumables. The method is nor directly derived from USEPA based methods. The instrumentation and skill level required for the method is rated as medium. The time for analysis of one sample is approximately 55 minutes with an estimated sample rate of about 1 sample/hr.

LC-MS Methods for HAAs

LC-MS based methods are not widely employed in the conventional laboratories. This is largely due to the relatively high costs of purchasing and maintaining these instruments. As this technology become more stable and cheaper, LC-MS techniques will offer many attractive alternatives to traditional techniques. LC-MS instrumentation is commercially available, but the level of skill of operators is considered high compared to other, more traditional approaches such as GC, GC-MS, LC and IC. Despite the still limited availability of LC-MS, there are still several promising alternatives appearing in the literature for analysis of HAAs. These methods are generally not derived from USEPA based methods.

Hashimoto and Otsuki (1998) have described an ESI-LCMS method for the determination of HAA9 in water. HAAs were extracted from environmental samples with MTBE. The internal standard was 2,3-DCPA. A Waters 600E LC equipped with a Supelcogel C-610H HPLC column (7.8 mm i.d. × 30 cm long) was used. Negative ion ESI method with a TSQ-700 triple quadrupole MS was used to analyze samples. The injection volume was 25 µL with isocratic conditions. The mobile phase was 3% acetic acid in acetonitrile:water (20:80) with flow rate of 0.5 mL/min. The sheath gas pressure, capillary temperature and flow rate were 70-80 psi, 260 to 280°C and 0.3 mL/min to 0.7 mL/min. The electrospray needle voltage was operated at –4.5 kV. The MDLs for MCAA, DCAA TCAA, MBAA, DBAA, TBAA, BCAA, CDBAA, BDCAA were 0.029 µg/L, 0.003 µg/L, 0.070 µg/L, 0.027 µg/L, 0.014 µg/L, 0.070 µg/L, 0.009 µg/L, 0.069 µg/L and 0.059 µg/L, respectively. The percent recoveries of HAA9 averaged above 80%. The precision of the method varied from ± 2.2% to ± 7.1%.

Takino, Daishima, and Yamaguchi (2000) have used LC-ESI-MS and volatile ion pairing reagents to measure HAA concentrations in water. HAA9 concentrations were measured in a water sample spiked with the volatile ion pairing reagent dibutylamine and propan-2-ol was added as a mobile phase modifier.

The LC-MS was an HP 1100 series LC with a vacuum solvent degassing unit, a binary high pressure gradient pump, an automated sample injector, and a thermostatted column. An HP model 100 series diode array detector was connected in series with the MS detector. The LC column was Inertsil ODS3 (150 mm × 2.1 mm i.d.). The separation mode was a linear gradient elution for 20 minutes with a mobile phase of acetonitrile-water (5 mM) adjusted to pH 7 with acetic acid. The flow rate was 200 µL/min. After the eluent passed the diode array detector, 2-propanol was added at a flow rate of 50 µL/min using a tee and an isocratic pump. The injection volume was 500 µL. The MS was an HP 1100 series MSD single quadrupole with orthogonal spray ESI for the MS introduction. The nebulizer gas was N_2 at 350°C from a N_2 generator. The gas and fragmentor voltage for the ion source were optimized using the column with a commercially available HAAs standard (100 mg/L). The nebulizer gas was set at 60 psi, the drying gas flow was 10 L/min, and the fragmentor voltage was 40 V. Skimmer 1, skimmer 2 and the entrance lens were optimized using a calibrant delivery system and calibration standard (0.1 mL/min) set to 23 V, 47 V and 57 V, respectively.

The MSD was operated in negative ion mode. The surface tension of the column effluent was decreased by adding 2-propanol to post-column. This facilitated the ionization of the HAAs and gave increased sensitivity of a factor of two. The scan range was from the m/z range 70 amu to 500 amu using a step size of 0.1 and a scan speed of 0.5 scan/s. The following selected ion were monitored: MCAA (93 amu), MBAA (127 amu), DCAA (139 amu), BCAA (173 amu), DBAA (217 amu), TCAA (117 amu), DCBAA (163 amu), DBCAA (207 amu), and TBAA (250 amu).

The dwell time was 250 ms to 500 ms per ion. The halogen isotopic ratios for all compounds were also monitored to confirm results.

HAA9 in the concentration range 0.1 ng/L to 100 mg/L were determined using the technique with time-scheduled SIM. The linearity was very good with r values of calibration curves greater than 0.999. The MDL estimates ranged for MCAA, MBAA, DCAA, BDCAA, DBAA, TCAA, DCBAA, DBCAA and TBAA were 89 ng/L, 96 ng/L, 24 ng/L, 88 ng/L, 91 ng/L, 83 ng/L, 113 ng/L, 74 ng/L and 118 ng/L, respectively. The % RSD calculated from five replicant measurements of a 0.2 ng/L standard were as follows: MCAA (4.5%), MBAA (5.3%), DCAA (1.5%) BDCAA (4.2%), DBAA (6.1%), TCAA (4.9%), DCBAA (9.4%), DBCAA (5.2%) and TBAA (9.1%), averaging about 6% for all HAAs studied.

Magnuson and Kelty (2000) described a method for the analysis of all HAAs by LLE by using complexation agent to form stable association complex with detection by ESI-MS. The goal of forming complexes was to increase the mass of the ion used to quantitate the HAA by complexing with perfluoroheptanoic acid. The MDLs for the HAAs were less than 1 μg/L (0.13 to 0.64 μg/L). Standard addition method was used to quantify the HAAs to avoid the matrix effects. The injection loop was 200 μL with a Rheodyne model 7725 injector. The mass spectrometer was Finnigan MAT TSQ-700 equipped with a Finnigan electrospray interface. Spectra was acquired in the negative-ion mode by scanning the third MS over appropriate mass ranges. The scan time was 0.25 s and the applied ESI spray potential was 4.0 kV. The interface capillary temperature was 130°C. The sheath gas (nebulizer) pressure was 70 psi. The flow injection mode was at 30 mL/min injection volume. The carrier liquid was MeOH at 0.3 mL/min and the extraction solvent was MTBE.

Loos and Barcelo (2001) described a method for the determination of nine HAAs in water based on SPE enrichment followed by ESI-LC-MS in the negative ionization mode. They found that the LiChrolut EN is the best polymeric material for SPE sorbents. For complete LC separation of all compounds by ion pair chromatography they used triethylamine as volatile ion-pairing reagent. LC separations were performed with an HP 1090A system coupled in series with an HP 1100 MS equipped with ESI interface and a standard atmospheric pressure ionization source using ESI in the negative mode. The column was a LiChrospher RP-C_{18} (250 mm × 4 mm). The eluent was acetonitrile and water containing 5 mM triethyl amine and 5 mM acetic acid gradient (85% to 50% water in 12 minutes) at a flow rate of 0.5 mL/min. The drying gas temperature was 350°C and the nebulizer pressure was 55 psi. The vaporizer temperature was 400°C. The capillary potential was 5000 V and fragmentation potential 40 V. The MDL values were below the μg/L range and analysis time was shorter than typical GC-MS based techniques, but the cost and maintenance of the instrument was relatively expensive.

Kuklenyik, Ashley and Calafat (2002) developed a rugged, high-throughput (~15 min/sample and ~50 samples/day), sensitive and selective method for measurement of trace levels of TCAA in human urine. The method used SPE cleanup followed by isotope dilution negative ion LC-ESI-MS-MS. The instrument used was Agilent 1100 LC coupled with an API 3000 triple quadrupole MS. The LC column was Prism RP (2 mm × 100 mm × 3 μm f.t.) that was preceded by a 2 mm guard column (3 μm). The eluent was isocratic 75% MeOH/25% 5 mM ammonium acetate (pH 5.2) with a flow rate of 0.2 mL/min at ambient temperature. To form the negatively charged ions negative ion turbo ion spray source was used via a 2 mm i.d. column at 0.2 mL/min flow rate without postcolumn splitting. The MDL was 0.5 ng/mL. The mean % recovery varied from 97% to 124% and the % RSD was 8.5%. The technique is not labor intensive since sample preparation and analysis can be performed in ~15 minutes but the cost of the instrument is relatively high.

Roehl et al. (2002) described the application of ESI-IC-MS for the determination of HAAs in various water samples. The instrument used was Dionex DX-500 microbore IC, which consisted of a gradient pump, chromatography compartment and electrochemical detector. Analytical columns were Dionex IonPac AS9-HC and AS16 (250 mm × 2 mm i.d.) with respective guard columns AG9-HC and AG16 (50 mm × 2 mm i.d.) fitted with a Finnigan AQA single-stage quadrupole MS. To minimize the baseline noise in the conductivity detector the MS was coupled to IC through a grounded stainless steel union. HAAs were separated and detected at low µg/L using on-line sample preconcentration prior to an anion exchange separation with gradient elution. The eluent was 5 mM to 70 mM NaOH, flow rate 0.25 mL/min. The ESI probe was held at 275°C and -2.5 kV. The source CID voltage was at 10 V to 20 V, at negative detection ESI-MS. The method application for HAAs analysis is yet to be optimized and the cost of the instrument would be relatively expensive and require high level of operator skills.

Capillary Electrophoresis Methods for HAAs

Martinez, Borrull and Calull (1999b) reviewed existing capillary electrophoresis (CE) techniques for the analysis of several compounds of environmental interest. With each, a suitable system has to be used to enhance the sensitivity. With CE very low concentrations of compounds have been measured (µg/L levels) for HAAs. There are several variations of CE; including capillary zone electrophoresis (CZE), micellar electrokinetic capillary chromatography (MEKC), and capillary electrochromatography (CEC). Compared to chromatography, CE gives better resolution with lower solvent cost. The narrower diameters of the separation columns lead to smaller sample volumes, and ultimately fundamental detection problems arose because of the limited path lengths. There have been many studies that attempted to improve sensitivity. Some have included improving sensitivity by using various sample introduction techniques. Two injection modes are possible based upon hydrodynamic or electrokinetic principles. Lower MDLs can be achieved when the injection time increases, increasing at the same time the length of the sample plug injected into the capillary. However, if this length gets too large, it will negatively influence the separation efficiency. With inorganic metal cations altering the injection mode from hydrostatic to electrokinetic injection can enhance the signal. This type of injection is not suitable for determination of inorganic anions in rainwater solutions, unless an internal standard is used. The accuracy and precision of electrokinetic injection may be dependent upon several factors including: the number of analyses per vial, the cleaning of the capillary and electrodes with water, and the addition of an electromigrative additive. Electrokinetic injection has been successfully applied for HAAs in drinking water at the µg/L level. The total analysis times range from five to ten minutes for most methods.

The most common detectors used in CE are UV detectors. There are different ways to increase the effective pathlength across the capillary. For example, rectangular tubing can be used with optical pathlengths up to one mm. Another way is to use capillaries with bubble cells or use mirrors to reflect the incident light inside the capillary prior to detection.

On-line SPE preconcentration with CE has given good results. Off-line preconcentration can be performed by LLE, SPE, or SPME. CE is a promising method for analyzing HAAs in water. Perhaps the major advantage of these methods is the fast analysis time. Sensitivity can be improved by using SPME but this research is in its infancy. The operator skills must be relatively high for CE analysis at the current state of the art.

Hozalski and Zhang (2002) explored CE methods for HAA analysis. They studied the effects of buffer composition and sample injection with "on-line" and "off-line" sample preconcentration in HAA analysis. Their results were compared to those obtained using USEPA Method 552.2 (1995d).

Three different buffers were evaluated for HAAs separation. They included a phosphate buffer, a citric acid/lithium hydroxide buffer and 2, 6-napthalenedicarboxylic acid (NDC). The pH of the buffers ranged from 3.4-10. Four different modifiers were used to reverse the electroosmotic flow (EOF). The EOF modifiers studied included tetradecyltrimethyl ammonium bromide (TTABr), CTAB, hexadimethrine bromide (HDMBr) and diethylenetriamine (DETA). Comparisons were made between a straight capillary and a bubble cell capillary with pressure injection at 200 mbar. Trifluoroacetic acid (TFAA) was used as an internal standard. The phosphate and citric acid/lithium hydroxide buffers were investigated using direct UV detection at 191 nm at 25°C. The method using the NDC buffer used indirect UV detection at 235 nm.

Two of the CE buffer systems gave complete resolution and good peak shapes. The first buffer was 4 mM NDC and 0.5 mM CTAB at pH 7.5 (NDC buffer) and the second buffer was 12.5 mM NaH_2PO_4, 12.5 mM Na_2HPO_4 and 1 mM TTAOH at pH 5.7 (phosphate buffer). The phosphate buffer gave a stable baseline. The NDC buffer baseline was not as stable, leading the researchers to choose the phosphate buffer in their method. A straight capillary was compared to the bubble cell capillary using phosphate buffer with pressure injection at 600 mbar. The MDLs for HAAs ranged from 170 to 270 μg/L.

Preconcentration techniques were explored to lower (improve) the MDL estimates. One approach "on-capillary concentration techniques" used capillary isotachophoresis and electro kinetic injection. Another "off-line" approach used SPE with Lichrolut EN columns containing highly cross-linked styrene divinyl benzene resin. Electro kinetic injection used an applied voltage to load sample into the capillary via electrostatic attraction and significantly improved the sensitivity for the HAAs, but the calibration curves were non-linear and not reproducible when the sample matrix contained ions other than the HAAs. SPE using Lichrolut EN columns was successful at enriching HAAs, and mean % recoveries ranged from 75% to 99%. The MDLs for the SPE method ranged from 1.7 to 2.5 μg/L. Matrix effects were not as problematic when pressure injection was used. Further studies of the SPE method showed that the presence of NOM did not affect the mean % recoveries. The proposed CE method for HAA analysis used the phosphate buffer, a bubble cell capillary (75 μm i.d. × 56 cm e.l.), pressure injection (50 mbar for 12 s) and the detection wavelength of 191 nm.

The major disadvantage noted is that it took approximately five hours to process each sample. However, sample throughput may be improved by using a multi-port vacuum manifold for SPE preparation and therefore process several samples simultaneously.

Martinez, Borrull and Calull (1998b) compared CE and CZE for the analysis of HAAs in drinking water. Four different sorbents were studied for SPE followed by indirect UV detection. These include: a quaternary ammonium anion exchanger (LC-SAX), a highly cross linked polymer of styrene-divinylbenzene (LiChrolut EN), a graphitized carbon black (Envi-Carb) and a macroporous poly(divinylbenzene-co-N-vinylpyrrolidine) coplymer (Oasis HLB).

The CE instrument was an HP with diode array detector. The separations were carried out using uncoated fused silica capillary tubing (64.5 cm × 75 μm i.d.). When a new capillary was used, the capillary was washed for 60 minutes with hydroxide solution, followed by 60 minutes with deionized water and finally, 30 minutes with the running buffer. The detection was set at 235 nm in the indirect mode and the temperature was kept constant at 25°C.

The HAAs that were studied included MCAA, MBAA, DCAA, DBAA and TCAA. NDC was used as the electrolyte system. CTAB was used as EOF modifier. NaOH was used to adjust the electrolyte pH. The electrophoretic conditions and system operation were as follows: Separations were carried out by rinsing the capillary for three minutes with a background electrolyte immediately before the injection. The detector was set at 235 nm. The capillary temperature was kept constant at 25°C. A linear gradient was set at –15 KV for 0.5 minutes and finally at the same potential for a further three minutes because an increase of the sharp shape of TCAA was observed.

Off-line trace enrichment was carried out using four different commercial SPE cartridges. The only exception was when LC-SAX was used as a sorbent, because in this case the ionic form was required. The extractions were carried out using the Bond Elut/Vac Elut system. The samples were passed through the cartridges at a flow rate of approximately 15 mL/min. All samples and electrolyte systems were filtered through a 0.45 μm filter before preconcentration.

The electrophoretic separation of the HAAs was focused on the electrolyte concentration, pH and various EOF modifiers. In optimum conditions these compounds were separated in less than eight minutes. The linearity of response by hydrodynamic injection (40 mbar for 2 s), using standards prepared in MeOH-water (50:50,v/v) was found to be good between 1 mg/L and 30 mg/L. The MDLs for MCAA, MBAA, DCAA, DBAA and TCAA were 5 μg/L, 5 μg/L, 2 μg/L, 3 μg/L and 2 μg/L, respectively. The % RSD values for MCAA, MBAA, DCAA, DBAA and TCAA were 9.8%, 6.3%, 5.5%, 7.4% and 6.7%, correspondingly. The r values showed good linearity over the concentration range from approximately 5 to 80 μg/L.

Kim et al. (2001) described a method based on LLE followed by CZE for the determination of five HAAs with direct UV detection at 185 nm in tap water. They used a Unicam Crystal CE system equipped with UV detector to perform the CZE. Separation was carried out using a capillary column (75 μm i.d. × 365 μm o.d. × 93 cm f.t.) from Supelco. The carrier electrolyte was 25 mM phosphate 0.5 mM CTAC adjusted to pH 7.0 at 25°C. Hydrodynamic injection was performed at the cathodic side at 50 mbar and separation was carried out at –20 kV constant voltage with a current of 20 μA. The mean % recoveries ranged from 89% to 97% and the MDL for HAA5 was ~5 μg/L. To lower the MDL samples were concentrated using LLE by a factor of 2000 times. The analysis time was within seven minutes except for TCAA.

Martinez, Borrull and Calull (1999a) used CE for HAAs analysis. CE can give high-resolution separation in many cases without derivatization and this separation is generally faster. The researchers used CZE and studied different electrolyte systems and electrophoretic behavior of HAAs.

The CE instrument was an HP Model CE equipped with a UV detector and HP Chemstation chromatographic data system. Uncoated fused silica capillary tubing was used for CE. The HAAs evaluated included MCAA, MBAA, DCAA, DBAA and TCAA. Four different electrolyte systems were studied: (1) a pH 6, 12 mM potassium hydrogen phthalate buffer, (2) a pH 7.5, 4 mM naphthalene dicarboxylic acid dipotassium salt (NDC) buffer, (3) a pH 9.6 buffer solution of 20 mM borate and (4) a pH 8.7, 10 mM chromate buffer. In order to reverse electro-osmotic flow, the effect of adding 0.5 mm of hexadecyltrimethylammonium bromide (CTAB) as an EOF modifier in all electrolyte systems was also studied.

The best results were achieved when an electrolyte system consisted of 4 mM NDC and 0.5 mm CTAB. Concentration range was 2-30 mg/L and exhibited good linearity (r>0.99). The MDL was in the low mg/L concentration levels. The % RSD ranged from 6.6% to 14.5%. The mean % recoveries ranged from 54% to 83%, but were generally in the 50% range.

In another study, Martinez et al. (1998) evaluated whether CZE was suitable for separating the HAAs as an alternative to the USEPA method for determining these compounds. A similar CE instrument was used as the one described in the previous study. Two different electrolytes were studied during this research: potassium hydrogen phthalate (phthalate) and 2,6-naphthalenedicarboxylic acid dipotassium (NDC). CTBA and tetradecyltrimethylammonium bromide (TTAB) were used as the EOF modifiers. LLE was used to carry out the off-line trace enrichment process. The absorbance was monitored at 254 nm and 235 nm for the phthalate and NDC electrolyte respectively. NDC and phthalate were used to analyze HAAs because the ionic mobilities of the carrier electrolytes and the sample ions mismatched. To make noise smaller and indirect absorbance stronger, particular concentration of electrolyte was chosen. NDC and phthalate electrolytes gave very similar results, where % RSD estimates ranged from 1.1% for MCAA to 3.4% for DBAA in the phthalate electrolyte and from 1.1% for MCAA to 4.2% for BCAA in the NDC electrolyte.

The NDC method was used to measure the concentrations of HAAs in drinking water. Calibration curves were prepared for standard solutions in the concentration range from 10 to 60 μg/L. The results showed that r values were greater than 0.999 for all compounds but MCAA, for this compound the r value was 0.989. Higher concentrations gave more poorly resolved peaks. The MDL estimates ranged from 3 μg/L for DCAA to 5 μg/L for the other HAAs. The results showed that the NDC method provides better sensitivity and selectivity for HAAs than phthalate electrolyte. The eight minute analysis time is attractive along with the fact that derivatization is not required. However the method still requires LLE as a preconcentration step.

Ahrer and Buchberger (1999) examined a new method based on CE coupled with atmospheric pressure ionization MS (CE-MS) as an alternative to the existing methods to analyze nine HAAs in water samples. MS detection was performed in quadrupole system in the negative SIM mode (HP 5989 B using pneumatically assisted ESI (HP 59987) interfaced with a CE-probe and a RF-only hexapole). The CE capillary was 70 cm length with 50 μm i.d. fused silica, and carrier electrolyte was 10 mM ammonium acetate in MeOH containing 12% acetic acid. The sheath flow rate for MS was 4 μL/min. using a solution of 2-propanol/water (80:20 v/v with 0.1% triethylamine). This solution was delivered using a syringe pump. The drying gas was N_2 at a flow rate of 2 mL/min. The drying gas temperature and MS temperature were 150°C and 100°C, respectively.

The MDLs ranged from 0.3 to 7.6 μg/L with r values of 0.995 to 0.999. Mean % recoveries were between 69% to 94% except for MCAA and MBAA where recoveries were in the range 33% to 36%. One of the limitations of this method is that interfacing CE with MS is more complex compared to LC-MS because the flow rate of CE carrier electrolyte is usually in the range of nL/min. This is why sheath flow (some μL/min) is necessary. The sensitivity and stability of MS detection depends highly on the solvents and the sheath liquid flow rate.

Xie, Zhou and Romano (2000) are developing methods for determination of HAAs in drinking water by CE. HAAs were analyzed in commercial standard (100 μg/mL of each HAA with MTBE) and field samples collected at Hershey, PA Water Treatment., A solution of ammonium chloride was added to all field samples to quench free chlorine residual. The researchers used Waters Quanta 4000 CE system. The instrument consisted of a negative power supply, a mercury lamp, a Waters Accu-Sep polyamide coated capillary (60 cm × 75 μm i.d.), an UV absorbance detector, and a Waters Millennium Chromatography Manager data system. A solution of 25 mM phosphate and 1.0 mM tetradecyltrimethyl ammonium hydroxide (TTAOH) was used as the electrolyte solutions. HAA species were detected with direct UV at 185 nm.

A standard curve was in range from 4 to 100 μg/L for each HAA species. The MDL values were less than 3 μg/L for mono-HAAs and di-HAAs, and ranged from 4.2 to 7.0 μg/L for tri-HAAs. The spiking recovery for eight HAAs (except TBAA) was in the range from 74.8% to 107%. Good linearity was achieved and r values were in the range from 0.991 to 1.000.

UV detection was not sensitive enough and samples were pre-concentrated approximately 1000 times using LLE and back extraction. Both extractions took about 30 to 45 minutes and were considered to be one of the disadvantages of this method. Researchers are working to decrease extraction time by trying to automate LLE. They are also trying to improve peak asymmetry and method sensitivity. This method can be used for analysis of HAAs after improvements are done. A CE instrument was relatively inexpensive but did require a qualified operator. The total analysis time was about one hour.

Soga and Imaizumi (2001) described the broad application of CE methods for the analysis of inorganic anions, organic acids, amino acids, nucleotides, carbohydrates and other anionic compounds. Indirect UV detection was suitable for HAA analysis. The CE instrument was Agilent CE system equipped with a diode array detector. Separations were made on fused silica capillaries with 50 μm i.d. × 112.5 cm length. The electrolyte solution used was 20 mM 2,6-pyridinedicarboxylic acid (PDC) and 0.5 mM cetyltrimethylammonium hydroxide (CTAH) was used as EOF modifier.

In summary, the CE based methods show promise for the analysis of HAAs in drinking water distribution systems. They offer particularly attractive separation capabilities in a relatively short time, but improvements in detection are needed to avoid sample preparation and preconcentration steps. At present, the technique is not widely applied in the drinking water community. The instruments are available commercially. The level of instrumentation and skill level of the analysts would range from medium to high. At present, these methods are not derived from USEPA based methods. More research needs to be done in the area of automating the sampling steps in CE to allow for on-line monitoring.

Colorimetric Based Methods for HAAs

There have been several colorimetric methods for TCAA based on the Fujiwara chemistry discussed earlier for THM analysis. This chemistry can be adapted for the analysis of TCAA. For example, Ogata, Takatsuka and Tomokuni (1970) developed a colorimetric method based on the Fujiwara chemistry for measuring TCAA concentrations in urine. As with other methods of this type, the chemicals that were required are commercially available and the level of equipment and instrumentation is relatively low—standard volumetric glassware and a UV visible spectrophotometer. The absorbance maxima for TCAA were 530 nm and 370 nm. The mean % recoveries for TCAA ranged from 93% to 96%. A clear disadvantage of the method is that it does not detect monohalogenated species such as MCAA and MBAA, but the method could be readily automated for on-line monitoring using flow injection analysis. Bhattacharjee, Cherian and Gupta (1991) proposed a Fujiwara based method for TCAA that was based on the reaction of TCAA with pyridine in basic solution with mild heating to form a red-colored chromophore. This red color was discharged by addition of a few drops of acetic acid and a mixture of sulfanilic and formic acid to obtain a species that absorbs at 505 nm. The reproducibility of the method was evaluated at the 5 mg/L level. The % RSD was about 3%. The method did not exhibit significant interferences. While detailed MDL, accuracy and precision studies were not reported, the analysis time was promising at approximately 15 minutes per sample. More recently, Pillai, Rastogi and Gupta

(1999) proposed a modified version of the Fujiwara chemistry that is based on measuring the absorbance produced when polyhalogenated compounds like TCAA react with pyridine in the basic solution. The absorption spectra of the dye formed showed a maximum absorbance at 520 nm, where the reagent blank showed negligible absorbance. The % RSD was good at the 1 mg/L in water. The mean % recoveries were found typically greater than 99%.

There have been colorimetric methods based upon other chemistries reported for TCAA analysis as well. Many of these methods were developed for analysis of TCAA in urine. For example, Tanaka and Ikeda (1968) developed a colorimetric method that used oxidation by chromium trioxide in water with concentrated nitric acid. A visible spectrometer was used for all absorbance measurements. Rathore, Saxena and Begum (1992) reported a selective fluorescence spot test for the detection of TCAA in soil and water. In this method, filter paper was soaked in the reagent solutions to prepare a test strip that was used to detect TCAA. The paper was exposed to UV light at 366 nm to develop the color. Aqueous and ethanolic solutions containing TCAA were mixed with NaOH solution and phenol solution in a small test tube. The solution was added to the test paper, dried and UV light (366 nm). A greenish yellow stain on a purple background indicated the presence of TCAA. Analysis was done on tap and river water samples. The lower detection limit for TCAA in drinking water was approximately 50 μg. This method shows promise as a simple field based method for TCAA. Pal (1996) proposed a colorimetric method for measuring TCAA concentrations in water based on measuring the absorbance produced in the photo-activated reaction with diphenylamine in acidic medium. Absorbance measurements were done with a typical UV-visible spectrophotometer. A portable germicidal lamp was used or color development. The absorption spectra of the green dye formed, showed a maximum absorbance at 660 nm. Good linearity was obtained in the concentration range up to 640 mg/L for TCAA. The % RSD for the method was 7%. A disadvantage of the method is that the photolysis was carried out in acetone, The photolysis time was about 30 minutes, thus limiting the sample throughput. A typical analysis time is estimated at about 40 minutes.

In summary, the colorimetric methods offer some attractive alternatives for HAA analysis. These methods can all be carried out with relatively low level of instrumentation and analyst skills. The materials and consumables are largely commercially available. Applied alone, these methods will likely be limited to measurements of total concentrations of HAAs and will not include monohalogenated species like MCAA and MBAA. Another possible application would be to use these colorimetric based methods in conjunction with IC methods and conduct post-chromatographic column derivatization to increase sensitivity. These methods are not derived from USEPA methods.

Electrospray Ionization-High-Field Asymmetric Waveform Ion Mobility Spectrometry-MS

Ells et al. (1999) reported a method for measuring brominated and chlorinated DBPs using electrospray ionization, high-field asymmetric waveform ion mobility spectrometry and MS (ESI-FAIMS-MS). Preliminary research with this method showed promise for analysis of six HAAs. In these studies, the MDL values ranged from 0.5 ng/mL to 4 ng/mL with no preconcentration, derivatization or chromatographic separation prior to analysis. In subsequent studies, Ells et al. (2000) used ESI-FAIMS-MS to measure the concentrations of nine HAAs at ng/L levels in source and treated drinking water. The following MDL values were reported: MCAA (28 ng/L), DCAA (10 ng/L), TCAA (36 ng/L), MBAA (18 ng/L), DBAA (14 ng/L), TBAA (6 ng/L), BCAA (27 ng/L), BDCAA (13 ng/L), CDBAA (5 ng/L). The method showed good linearity for HAA9.

More recently, Gabryelski, Wu and Froese (2003) developed standard addition ESI-FAIMS-MS for the measuring the concentrations of HAAs in drinking water and compared the quantitative results to conventional GC based methods. The MDL values were as follows: MCAA (0.31 μg/L), DCAA (0.17 μg/L), TCAA (0.18 μg/L), BCAA (0.07 μg/L), BDCA (0.05 μg/L), DBAA (0.07 μg/L), MBAA (0.12 μg/L).

The ESI-FAIMS-MS method is interesting for several reasons. While it is not well established compared to other more traditional instrumental analysis techniques, no sample preconcentration is necessary. The method is fast, relatively simple and shows good sensitivity for a newly developing technique. There is a significant reduction in the chemical background compared to typical ESI-MS methods. In fact, chromatographic separation is eliminated. However, the instrumentation was not readily available for construction of such a routine analyzer and the level of operator error is expected to be quite advanced at this stage in its development.

Other Notable Sample Preparation/Introduction Methods

There are some other notable sample preparation/introduction methods that are still in the development stage. These include the sequential injection liquid-liquid extraction (SILEX) system, the "Mixmaster" automated sample preparation system, and a sequential injection analysis (SIA) analyzer that uses restricted-access materials (RAM) for on-line monitoring.

The SILEX analyzer is reported to interface with both GC and LC instruments. The SILEX system can be used to automate LLE and derivatization steps that are required for HAA analysis (Wolcott and Olsen in review). A typical SILLEX configuration might be comprised of a ten-port selection valve and a Cavro syringe pump. The resulting instrument could be used to chemically condition the sample, expose the sample to extracting solvent, mix the sample and solvent to allow extraction occur, and separate the sample and solvent immiscible zones by means of a separation cell. The solvent zone (containing the analyte) can then be introduced to a GC or HPLC. The SILEX analyzer eliminates the need for dynamic phase separation. It allows for online pre-extraction and chemical conditioning. It reduces solvent, reagent and sample usage and waste. Moreover, all off the steps are completely automated. The SILEX system has not been used for analysis of drinking water DBPs and more research needs to be done in this area.

Vogt et al. (2003) are developing an automated sample preparation system—the "mixmaster" to use in VOC analysis. A HS-free mixing system was designed using a piston pump, stainless steel tubes, and airtight glass syringes holding the stock solutions. The researchers used this system to investigate polymer or sol gel coated chemical sensors based on attenuated total reflection (ATR) spectroscopy. The sample preparation system worked on similar principles as SIA. Chemometric algorithms were used to calibrate the instrument. The components of the device are commercially available. For handling the mixing system a program named "MixMaster" was developed. At present, the mixmaster system is still being developed, but ultimately it may be applied to analysis of drinking water DBPs.

Satinsky et al. (2003) proposed a fully automated on-line system by coupling of solid phase extraction and restricted-access materials (RAM) with SIA. This SIA-RAM technique was used for drug analysis in blood serum, and also interfaced to HPLC. On-line sample preparation was performed on using a RAM alkyl-diol silica (ADS) column packed with LiChrospher RP-18 ADS 25 μm, silica pore diameter 60A. Detection was accomplished using a UV-visible spectrophotometer over the wavelength region 240 nm to 300 nm. The total analysis time for SIA-RAM was less than ten minutes. The cost of analysis was estimated to be relatively low compared to conventional analytical methods. While the SIA-RAM technique has not been applied to drinking water analysis, it does show promise as an on-line monitoring tool. Further research needs to be done before it could be used for THM and HAA analysis.

SUMMARY OF HAA METHODS

There are many papers in the literature pertaining to analysis of HAAs. A wider variety of techniques are used compared to THMs analysis. There is no clear agreement on which class of techniques offer advantages. The LC and LC-MS based techniques seem to hold great promise for minimizing sample preparation steps and analyzing several HAAs. A compilation of the MDL, accuracy and precision data as well as other characteristics is presented in Table 4.2 at the end of the report.

Table 3.1
Expected MDL, accuracy and precision results for HAAs using USEPA Methods

Method	Chemical species	MDL (μg/L)	Mean % recovery	% RSD
USEPA 552 LLE,	MCAA	0.052	74	38
Derivatization, GC-ECD	MBAA	0.074	58	7
	DCAA	0.015	84	10
	TCAA	0.085	78	59
	DBAA	0.015	58	14
USEPA 552.1* LLE,	MCAA	0.21	98	4.6
Derivatization, GC-ECD	MBAA	0.24	73	7.9
	DCAA	0.45	110	7.7
	TCAA	0.07	60	4.0
	DBAA	0.09	58	6.4
USEPA 552.2 LLE,	MCAA	0.273	94.7	7.3
Derivatization, GC-ECD	MBAA	0.204	102	5.0
	DCAA	0.242	84.7	9.6
	TCAA	0.079	93.0	10
	DBAA	0.066	95.4	9.2
USEPA 552.3 Micro-LLE,	MCAA	0.17 / 0.20	95.8 / 81.4	4.0 / 5.1
Derivatization, GC-ECD	MBAA	0.027 / 0.13	92.2 / 90.7	1.8 / 3.7
(MTBE/TAME)	DCAA	0.020 / 0.084	93.8 / 97.8	1.6 / 2.2
	TCAA	0.019 / 0.024	105 / 107	0.52 / 0.90
	DBAA	0.012 / 0.021	105 / 105	0.63 / 0.86

* USEPA Method 552.1 exhibits known matrix problems

CHAPTER 4
RECOMMENDATIONS FOR FURTHER RESEARCH

CONCLUSIONS FOR THE FINAL REPORT

The literature of analytical chemistry contains a wealth of information pertaining to the analysis of THMs and HAAs in drinking water. A large number of papers have been reviewed and summarized in this Report. The focus was on collecting data about the specific operation of the instruments and the sampling, sample preparation steps that were used so that these methods could be evaluated as to their adaptability for on-line real-time monitoring in drinking water distribution systems. Several of the reviewed methods show promise and include a wide array of methodologies ranging from simple colorimetric spot tests to more complex LC-MS methods and beyond. Careful review of the literature continues in our laboratory.

An electronic database has been submitted in rich text format (RTF) that is searchable via the "find" feature of most commercially available word processors. This database includes references to all THMs and HAAs papers reviewed to date. It is the hope of the authors that this database will be a valuable resource for the AwwaRF members and researchers interested in analytical method development. Such a database is relatively easy to update and distribute to subscribers, subscribing utilities and other interested parties.

As has been discussed earlier in the Report, the methods that are applied to monitoring THMs and HAAs are very different. This difference stems largely from the difference in chemical properties exhibited by THMs and HAAs, specifically polarity and volatility. There are two tables designed to summarize the results of this comprehensive literature review. Because of their large size (the tables require several pages) and to provide easy access for the reader, these tables have been added at the end of the Chapter. Table 4.1 summarizes the methods reviewed for THMs analysis. Table 4.2 summarizes the papers reviewed for HAAs analysis. The methods were evaluated based on several criteria. In addition to MDL, mean % recovery, and % RSD, the methods were evaluated based on the following criteria: commercial availability of instrumentation, whether it is based on an USEPA method, the level of instrumentation, estimated analytical skill level of analyst, the instrumentation cost, the ruggedness and portability of instrumentation, the analysis time, the sampling rate, the ability to adapt to on-line monitoring as a real or near real time method and possible automation modes. The necessary instrumentation and skill level were ranked as low, medium and high. Low level instrumentation includes wet chemical and spectrophotometric based methods. Medium level instrumentation refers to traditional instrumental based chromatography such as GC and HPLC with non-MS based detection. High level of instrumentation included hyphenated methods such as GC-MS and LC-MS as well as less well-established methods of analysis.

In addition, there are other tables presented to provide a resource to the readers that collect pertinent information related to the analysis. Table 4.3 lists the common commercially available GC columns that were used for THMs analysis in the methods that were reviewed. Table 4.4 lists the GC columns that were used for HAA analysis. Table 4.5 lists the common internal and surrogate standards that were used in the reviewed methods for THMs and HAAs. Table 4.6 lists the common HPLC / IC columns that were used for HAA analysis. All of these tables are presented at the end of this Chapter for easy access. Based upon the review, several conclusions about the methods that show promise for real-time monitoring have been drawn. However, before methods are reviewed with respect to application to monitoring, it would be helpful to explore the characteristics of a proposed "Ideal Method."

Defining the Ideal Method

The focus of the project is to ultimately develop real-time or near real-time methods for monitoring THMs and HAAs in drinking water distribution systems. Ideally, it would be desirable to develop one analytical method that would be capable of monitoring both THM and HAA concentrations in the distribution system. Realistically, it is unlikely that this will be possible given the differences in chemical properties that exist between these two classes of DBPs. An immediate and perhaps more obtainable compromise would be to have two Ideal Methods, one for THMs and one for HAAs. When dealing with complex samples such as drinking water, it has long been a goal of environmental analytical chemists to have available two methodologies, based upon different chemistries, for measuring concentrations in complex matrices. So ideally, it would be desirable to have two analytical methods based on two different chemistries for monitoring THMs and two different analytical methods based upon two different chemistries for monitoring HAAs.

From the standpoint of fundamental analytical chemistry, the MDL is a very important parameter for determining the reliability and applicability of an analytical method. For this particular study, the project advisory committee (PAC) and the authors agreed that the MDL is an important statistic. Contrary to typical trends in traditional analytical chemistry, specifically the desire to achieve lower and lower MDL values, the PAC and authors have agreed that the MDLs will be considered acceptable if it is at least 0.5 μg/L. However, as one reviewer notes, perhaps a more realistic MDL range would be in the range 0.5-5 μg/L. Two other parameters that are commonly used to compare analytical methods are accuracy and precision. For our purposes, the mean % recovery will be used to evaluate accuracy. Accuracy is defined as how close the value is to the true value. Precision, defined as the amount of scatter in the data, will be estimated using the % RSD. As a rule of thumb, the target goals of the mean % recovery will be 100% ± 5%. The % RSD will be deemed acceptable for monitoring studies if it is less than 10%. The method used to calculate the MDL, the mean % recovery and the % RSD will be based upon USEPA recommendations as described earlier in this report.

The PAC and authors agree that any analytical method developed for drinking water monitoring in distribution systems should not be adversely affected by the presence of natural organic matter. As a rule of thumb, it has been agreed that any method developed should provide acceptable results when total organic carbon (TOC) concentration is in the range 1-20 mg/L.

It is expected that the ideal analytical method or methods would provide data in real-time or near real-time. Real-time data collection typically implies that the analytical response will be instantaneous. The PAC and the authors have conceded that such instantaneous data collection, while certainly desirable, may be limited. For example such methods might provide only total THMs or total HAAs. While this information could be quite valuable, it is recognized that in many cases, it is desirable to be able to distinguish the different THMs and HAAs. To achieve this goal, it is likely that some sort of chromatography or other separation technique will be needed. This means that the data collection rate will be limited by the timescale of the separation process, for example the chromatogram. With this thought in mind, the PAC and the authors agree that in the ideal case it would be desirable to collect THM or HAA concentration data and have it available for decision-making in the water treatment plant within five to 20 minutes. However, in some cases it would be acceptable if the total analysis time gave a sampling rate of 1 sample/hr. So the

Ideal Method should be able to provide a sampling rate of at least one sample/hr—but the faster the better as long as other conditions such as MDL, accuracy and precision are met.

An additional characteristic of the ideal method would be that it could be accomplished with portable instrumentation and specifically designed for in the field monitoring. With respect to drinking water monitoring, it is assumed that electrical power will be available in the field. So perhaps a more appropriate way of thinking about field sampling would be carried out at monitoring station maintained throughout the system—or more specifically at different locations in the city. The Ideal Method would be able to be applied in such conditions.

It is expected that any methods and portable analyzers that are developed will be user friendly and that the skill level of the operator will likely not be at the typical level of the laboratory chemist with and associates or bachelor's degree—but significantly less specialized in analysis. With this in mind, an obvious path toward user friendliness would be for the method to be largely automated and have the ability to operate with minimal operator attention. The desire to have a simple automated method may seem to circumvent the well-established USEPA method for drinking water monitoring. This is not the case. Any portable automated methods that are developed should naturally be compared to those well-established methods that are proposed by the USEPA for drinking water monitoring. This means that the USEPA methods will be the standard by which the Ideal Method will be compared.

It is expected that the Ideal Method will be affordable—even inexpensive. Expense is a relative term. After all, what might be affordable for one utility might seem outrageous for another. With this in mind, it is desirable that an array of methods be developed that range from simple automated colorimetric tests to more advanced techniques that may involve some sort of automated chromatography or electrophoresis. With this in mind, the PAC and the authors have agreed that a suite of methods be explored. It is desired that the high end cost limit be about $20,000—a relatively inexpensive automated analyzer. In addition, techniques will be explored that could offer semi-automated analysis for as little as $3000 as well as techniques that will enable utilities to adapt existing equipment for automated or semi-automated monitoring. It is also desirable that the analysis time be as short a possible and require minimal calibration and maintenance. An additional attractive characteristic would be for the device to be able to interface to existing communication networks for early warning capabilities—perhaps through computer or cellular phone networks.

With these characteristics of the Ideal Method in mind, it should be noted that is unlikely that any one analytical method will meet all these requirements. However, it may be possible to develop several methods that will provide all the desired characteristics through overlapping capabilities.

Summary of Methods for THMs Analysis

There are 34 methods reviewed that are used for THMs analysis. Of these methods, 29, the vast majority, are GC based methods. This makes sense because the THMs are non-polar and volatile and can readily be analyzed by GC. The major difference that exists among these GC based methods is derived from the sample introduction technique. Most of the methods can be grouped into the following categories: LLE, P&T, HS, MI, or SPME.

Five of these methods including two USEPA methods used LLE to extract and preconcentrate the THMs before injection into the GC or GC-MS—typically via manual injection or automatic sampling using a robotic sample injection device. The problem with LLE is that the technique is largely dependent upon manual sample handling. In short—these methods are difficult to automate. Over the years there have been a number of approaches that have sought to automate LLE. Unfortunately, very few of these have been successfully applied to automated on-line monitoring. There are approaches that may show promise for accomplishing this goal, such as FIA or SIA techniques and their variants. However there are other options that offer simpler solutions to automating LLE for THMs and these techniques should be explored before much effort is devoted to automating the LLE process for THMs analysis.

Nine of the methods, including two USEPA methods, used P&T as a sample introduction method for introducing THMs into a GC or GC-MS. A large part of the attraction of P&T technologies is undoubtedly derived from the fact that P&T has been semi-automated. Several of the papers reviewed in this Report have been fully automated by some alteration in an existing system. These altered methods are very clever and provide acceptable results. However from the standpoint of the Ideal Method, many of them will exceed the price range of under $20K and few would be considered portable or even "luggable."

Two of the methods reviewed explore HS sample introduction methods for THMs. The HS approaches presented here exhibit problems with low precision (high % RSD). In addition, they are not readily adaptable to real-time monitoring. There are other techniques that show more promising results. Ultimately HS techniques are not recommended for further study.

Five of the methods involved MI derived techniques. MI techniques and their derivations of MI show promise as real-time monitoring sample introduction techniques. Many of these MI devices can be constructed with commercially available materials. The SCMS sampling probe is a commercially available and relatively affordable device. One probe costs approximately $1500—significantly less than a P&T sampler. The ruggedness and simplicity of this device make it attractive as a THM sampling device. It may be possible to construct a simple and portable SCMS-GC based device for well under the target price of $20K using components that are commercially available from reliable vendors. A rough estimate of one such device build from commercially available components and their approximate cost follow: Global FIA offers the SCMS probe ($1500), calibration vessel ($800) and flow cell sampling device ($400). Valco Instrument Company offers a heated valve enclosure ($200), a heated column enclosure ($300), two instrument temperature controllers ($770), a PDPID detector ($5400), a six-port injection valve with electrostatic actuator ($1200) and a serial valve interface ($610). Using these components in combination with a laptop computer ($3000) and a commercially available software package like Peak Simple ($3000) along with additional miscellaneous tubing and fittings ($300) could lead to the construction of a simple SCMS-GC analyzer for THM monitoring. These components combined total approximately $18K. A different SCMS-flow cell ($400) would be needed for each sampling site, but this could be an affordable option for many utilities. It is also possible that the SCMS-sampling probe cold be interfaced to the commercially available zNose fast GC with SAW detector. Such a system would cost a little more—approximately $20K, but include everything one would need with the exception of the SCMS probe and injection valve. A zNose based SCMS sampler would cost approximately $25K to construct but might provide additional advantages of less expensive operation costs, and networking capabilities.

Five of the methods explored SPME approaches to THM analysis. SPME offers many advantages for THM analysis. SPME is based on commercially available materials and relatively inexpensive—the initial cost of establishing SPME sampling capabilities in a water laboratory is $2K to $3K. However, a GC is still required to complete the analysis. While it would be difficult to completely automate the SPME sampling approach, it is possible to design experiments that would allow SPME sampling to provide data well within the one-hour time limit imposed by the

Ideal Method. In addition, it might be possible to sample using SPME at remote sites and then return the sample to the laboratory for analysis. An additional attractive characteristic of SPME is that it is relatively simple and non-analysts could be trained to sample the drinking water system through simple standard operating protocols. The SPME fibers are also reusable up to 100 times helping to reduce the cost of applying this method. So while not fully automated, the SPME approach provides many alternatives to increasing sample throughput and decreasing sample preparation time in THM analysis.

There are four colorimetric methods presented for THM analysis. Colorimetric methods offer attractive alternatives in drinking water. Perhaps the largest disadvantage is that they are often oversimplified and applied in situations where they provide unreliable results. However if care is take to understand the sample matrix and the possible interferences that might be present, the colorimetric approach can lead to good results at a relatively low cost. The analytical skill required is also relatively low. Perhaps the major advantage of colorimetric methods is that it is possible to completely automate them via FIA or SIA techniques or more advanced lab on a valve technology (LOV) or other microfluidic devices. FIA and SIA analyzers are relatively inexpensive compared to more advanced analytical techniques such as GC or GC-MS. A typical fully automated FIA or SIA analyzer can be purchased for well under $20K and can sample directly from the water distribution system continuously with minimal operator intervention. The more advanced capabilities such as LOV or microfluidic devices will eventually offer other advantages but at present are likely beyond the capabilities of the typical water utility at the present time.

In summary, there are several options available for real-time monitoring of THMs in drinking water distribution systems. These methods range from simple colorimetric methods and spot tests to intermediate methods that are derived from USEPA and similar chromatography methods and FIA and SIA technologies, to more advanced methods that are at the cutting edge of analytical chemistry including microfluidic devices and lab on a valve and lab on a chip applications.

Summary of Methods for HAAs Analysis

There are approximately 58 methods reviewed that are applied for HAA analysis. Of these methods, about 20 are GC based methods that require some sort of derivatization before the HAAs can be quantitated. These methods include the USEPA analytical methods for HAAs. These methods offer very good sensitivity, accuracy and precision. Unfortunately, the derivatization steps are often relatively complicated and require a great deal of manual manipulation of the sample. Some attractive options exist for automating the derivative steps and then introducing the samples automatically into the GC. For example, it is likely that FIA or SIA techniques such as SILEX could be applied to automating derivatization. Once converted to more volatile derivatives, GC or GC-MS analysis offers attractive options. However the chemical kinetics of the derivative processes will certainly fundamentally limit the sample throughput of such methods. It is unlikely that any of the HAA methods that require derivatization will provide data much faster than the 1sample/hr limit imposed by the Ideal Method. Nevertheless, these alternatives need to be explored.

SPME offers many simplifying options to traditional analysis. Perhaps the greatest advantage is that *in situ* techniques have been proposed that greatly simplify derivatization and sample introduction in GC. SPME and HS-SPME do seem limited when it comes to full automation. However as with the THM methods, SPME may provide attractive advantages. Even with derivatization, preliminary results show that it is possible to achieve the 1 sample/hr limit imposed by the Ideal Method at acceptable MDLs, accuracy and precision. While not fully automated, the

SPME approach may offer many alternatives to increasing sample throughput and decreasing sample preparation time in HAA analysis.

There are several colorimetric methods presented for HAAs analysis. These techniques appear to be limited to the di-halogenated and tri-halogenated species but they do offer attractive alternatives particularly when combined with FIA or SIA technologies that make possible complete automation. As discussed with the THMs, the largest disadvantage of simple colorimetric methods is that they are often oversimplified and applied to samples where they provide unreliable results. However, as long as precautions are taken to understand possible interferences that might be present, the colorimetric approach can lead to good results at a relatively low cost.

HPLC, IC and LC-MS options are very attractive for HAA analysis. The largest disadvantage is the cost associated with these instruments—particularly LC-MS. However lower end, older technologies that employ LC are becoming more routine and less expensive. Of the LC methods, IC techniques offer very attractive options. The major advantage of IC is that the majority of eluents that are used are aqueous based. This means that it is relative easy to automate sample injection with a typical six-port sample injection valve controlled by ValvePro, the shareware program available from the U of M laboratory. If preconcentration is needed, FIA or SIA can be applied to accomplish this goal. If the sensitivity of these methods needs to be enhanced, then techniques can be borrowed for FIA that allow for post-chromatographic column reaction with colorimetric reagents. This approach of combining IC and colorimetric methods can circumvent the limitations of each: increasing the sensitivity of detection and increasing the selectivity of the colorimetric methods by separating the HAA species via the chromatographic column. Another advantage of LC and IC methods is the wide array of commercially available consumables such as columns and pretreatment devices. It should also be noted that the automation of sample introduction in LC and IC is an investment in the future of drinking water analysis. Certainly any technique developed now for lower end LC and IC technologies will be readily adaptable to LC-MS methods as that technology become less expensive and thus more widely applied.

In summary, there are several options available for real-time monitoring of HAAs in drinking water distribution systems. While the analysis and monitoring problems presented by HAAs are typically more daunting than those presented by THMs, they can be overcome. These methods range from simple colorimetric methods and spot tests to intermediate methods that are derived from USEPA and similar chromatography methods, LC and IC methods, FIA and SIA technologies, and ultimately LC-MS approaches and even more advanced methods that are at the cutting edge of analytical chemistry.

Table 4.1
Summary of analytical methods from the literature for measuring THMs concentrations in drinking water

Analytical method (Reference)*	THM species	MDL (µg/L)	Mean % recovery	% RSD	Comm. avail. / EPA basis	Equipt. required / Est. skill level†	Instru. cost ($)‡ / Rugged and portable	Anal. time / Samp rate	Adapt to on-line monitoring / Real or near real time	Possible automation modes
USEPA 502.2 Purge and Trap-GC-PID-Hall (USEPA 1995a)	$CHCl_3$	0.02	98	2.5	YES / YES	Instrum. Method / Medium	20-24K / YES	50-60 min / 1-1.2 per hr	Readily Adaptable / Near-Real Time	Continuous flow Purge and Trap; SCMS-GC, SPME, HS-SPME
	$CHBrCl_2$	0.02	97	2.9						
	$CHBr_2Cl$	0.8	97	2.8						
	$CHBr_3$	1.6	106	5.2						
	TTHMs	2.4	100	3.4						
USEPA 524.2 Purge and Trap-GC-MS (USEPA 1995b)	$CHCl_3$	0.03	90	6.1	YES / YES	Instrum. Method / High	60-70K / YES§	50-60 min / 1-1.2 per hr	Readily Adaptable / Near Real Time	Continuous flow Purge and Trap; SCMS-GC, SPME, HS-SPME
	$CHBrCl_2$	0.08	95	6.1						
	$CHBr_2Cl$	0.05	92	7.0						
	$CHBr_3$	0.12	101	6.3						
	TTHMs	0.28	94.5	6.4						
USEPA 551 Liquid-liquid extraction GC-ECD (USEPA 1990a)	$CHCl_3$	0.002	80	2.4	YES / YES	Instrum. Method / Medium	14-15K / YES**	40-60 min / 1-1.5 per hr	Possibly Adaptable / Near Real Time	FIA or SILEX techniques
	$CHBrCl_2$	0.006	99	2.7						
	$CHBr_2Cl$	0.012	104	1.5						
	$CHBr_3$	0.012	109	0.7						
	TTHMs	0.032	98	1.8						
USEPA 551.1 Liquid-liquid extraction GC-ECD (USEPA 1995c)	$CHCl_3$	0.080	99	2.9	YES / YES	Instrum. Method / Medium	14-15K / YES**	40-60 min / 1-1.5 per hr	Possibly Adaptable / Near Real Time	FIA or SILLEX techniques
	$CHBrCl_2$	0.068	89	4.1						
	$CHBr_2Cl$	0.008	85	3.4						
	$CHBr_3$	0.020	96	2.8						
	TTHMs	0.176	92	3.3						

(continued)

Table 4.1 (Continued)

Analytical method (Reference)*	THM species	MDL (μg/L)	Mean % recovery	% RSD	Comm. avail. / EPA basis	Equipt. required / Est. skill level†	Instru. cost ($)‡ / Rugged and portable	Anal. time / Samp rate	Adapt to on-line monitoring / Real or near real time	Possible automation modes
Liquid-liquid Extraction-GC-ECD (Nikolaou et al. 2002b)	$CHCl_3$	0.010	100	3	YES / YES	Instrum. Anal. / Medium	14-15K / YES**	40-52 / 1.2-1.5 per hr	Possibly Adaptable / Near Real Time	FIA or SILLEX techniques
	$CHBrCl_2$	0.005	100	5						
	$CHBr_2Cl$	0.007	101	3						
	$CHBr_3$	0.010	102	3						
	TTHMs	0.032	101	3.5						
Purge and Trap-GC-ECD (Allonier et al. 2000)	$CHCl_3$	0.05	93-112	2-8	YES / YES	Instrum. Anal. / Med-High	22-24K / YES**	30-40 / 1.5-2 per hr	Readily Adaptable / Near Real Time	Continuous flow Purge and Trap; SCMS-GC, SPME, HS-SPME
	$CHBrCl_2$	0.02								
	$CHBr_2Cl$	0.02								
	$CHBr_3$	0.07								
	TTHMs	0.16								
Laboratory-built Purge and Trap device-GC-ECD (Zygmunt 1996)	$CHCl_3$	At the level of 1 ng/L	100	2.9-4.1	Combo of comm. and non-comm. / YES	Instrum. Anal. / High	22-24K / YES**	70-80 / 0.8 per hr	Readily Adaptable / Near Real Time	Continuous flow Purge and Trap; SCMS-GC, SPME, HS-SPME
	$CHBrCl_2$		97							
	$CHBr_2Cl$		91							
	$CHBr_3$		81							
	TTHMs		92							
Purge and Trap GC-MS (Nikolaou et al. 2002b)	$CHCl_3$	0.01	99	5	YES / YES	Instrum. Anal. / Med-High	68-70K / YES§	45-50 / 1.2 per hr	Readily Adaptable / Near Real Time	Continuous flow Purge and Trap; SCMS-GC, SPME, HS-SPME
	$CHBrCl_2$	0.05	103	7						
	$CHBr_2Cl$	0.01	95	5						
	$CHBr_3$	0.01	94	5						
	TTHMs	0.08	98	5.5						

(continued)

Table 4.1 (Continued)

Analytical method (Reference)*	THM species	MDL (µg/L)	Mean % recovery	% RSD	Comm. avail. / EPA basis	Equipt. required / Est. skill level†	Instru. cost ($)‡ / Rugged and portable	Anal. time / Samp rate	Adapt to on-line monitoring / Real or near real time	Possible automation modes
Continuous Flow Purge and Trap GC-MS (Chen and Her 2001)	$CHCl_3$	0.020	97.1-98.7	1.4-10.5	Combo of comm. and non-comm. / YES	Instru. Anal. / High	74-75K / YES§	5-10 min / 5-6 per hr	Readily Adaptable	Continuous flow Purge and Trap; SCMS-GC, SPME, HS-SPME
	$CHBrCl_2$	0.060								
	$CHBr_2Cl$	0.100								
	$CHBr_3$	0.120								
	TTHMs	0.3								
Purge and Trap-GC-ECD (Zhang, Qian and Yang 2001)	$CHCl_3$	Sub-ng/L conc. level	100	N/A	YES / YES	Instrum. Anal. / Med-High	22-24K / YES**	30-40 / 1.5-2 per hr	Readily Adaptable	Continuous flow Purge and Trap; SCMS-GC, SPME, HS-SPME
	$CHBrCl_2$		95							
	$CHBr_2Cl$		95							
	$CHBr_3$		71							
	TTHMs		90							
Purge and Trap-GC-MS (Zhang, Qian and Yang 2001)	$CHCl_3$	Sub-µg/L conc. level	100	N/A	YES / YES	Instrum. Anal. / Med-High	68-70K / YES§	30-40 / 1.5-2 per hr	Readily Adaptable	Continuous flow Purge and Trap; SCMS-GC, SPME, HS-SPME
	$CHBrCl_2$		95							
	$CHBr_2Cl$		95							
	$CHBr_3$		75							
	TTHMs		91.3							
HS Sampling-GC-ECD (Wang, Simmons and Deininger 1995)	$CHCl_3$	0.1	104	7.7	Combo of comm. and non-comm. / YES	Instrum. Anal. / Medium	20-22K / YES**	30-40 / 1.5-2 per hr	Possibly Adaptable	Automated sampler for GC, FIA / SIA possibilities
	$CHBrCl_2$	0.01	148	2.6						
	$CHBr_2Cl$	0.01	99	6.0						
	$CHBr_3$	0.01	85	8.5						
	TTHMs	0.13	109	6.2						

(continued)

Table 4.1 (Continued)

Analytical method (Reference)*	THM species	MDL (μg/L)	Mean % recovery	% RSD	Comm. avail. / EPA basis	Equipt. required / Est. skill level†	Instru. cost ($)‡ / Rugged and portable	Anal. time / Samp rate	Adapt to on-line monitoring / Real or near real time	Possible automation modes
HS Sampling-GC-ECD (Kuivinen and Johnsson 1999)	$CHCl_3$	<0.1	94	4.5	YES / YES	Instrum. Anal. / Medium	20-22K / YES**	53 min / 1.1 per hr	Possibly Adaptable / Near Real Time	Dynamic Headspace sampling
	$CHBrCl_2$	<0.1	96	3.6						
	$CHBr_2Cl$	<0.1	96	3.9						
	$CHBr_3$	0.2	100	10.5						
	TTHMs	<0.5	96	5.8						
SPME-GC-ECD (Chen, Mao and Hsu 1997)	$CHCl_3$, $CHBrCl_2$, $CHBr_2Cl$, $CHBr_3$, TTHMs	NR	112-121	4	YES / YES	Instrum. Anal. / Medium	17-18K / YES**	30-35 min / 1.7-2 per hr	Possibly Adaptable / Near Real Time	Automated sampler for GC, FIA / SIA possibilities
HS-SPME GC-MS (Cho, Kong and Oh 2003)	$CHCl_3$, $CHBrCl_2$, $CHBr_2Cl$, $CHBr_3$, TTHMs	0.005-0.01	NR	0.8-6.2	YES / YES	Instrum. Anal. / Med-High	60-61K / YES§	60-65 min / 0.9-1 per hr	Possibly Adaptable / Near Real Time	Automated sampler for GC, FIA / SIA possibilities
A simple membrane sampling device-GC-FID (Blanchard and Hardy 1984)	$CHCl_3$, $CHBrCl_2$, $CHBr_2Cl$, $CHBr_3$, TTHMs	NR	NR	NR	Combo of comm. and non-comm. / NO	Instrum. Anal. / Medium	14-15K / YES	24-25 min / 2.4–2.5 per hr	Readily Adaptable / Near Real Time	FIA / SIA techniques; Fast-loop Flow-cell

(continued)

Table 4.1 (Continued)

Analytical method (Reference)*	THM species	MDL (µg/L)	Mean % recovery	% RSD	Comm. avail. / EPA basis	Equipt. required / Est. skill level†	Instru. cost ($)‡ / Rugged and portable	Anal. time / Samp rate	Adapt to on-line monitoring / Real or near real time	Possible automation modes
SCMS-GC-ECD (Duty 2000)	$CHCl_3$	0.1	100	0.4	YES / YES	Instrum. Anal. / Medium	15-18K / YES**	28-30 min / 2-2.1 per hr	Readily Adaptable / Near Real Time	Injection valve, Fast loop-Flow Cell Arrangement
	$CHBrCl_2$	0.1	131	0.4						
	$CHBr_2Cl$	0.6	128	1.8						
	$CHBr_3$	6.1	120	12.8						
	TTHMs	6.9	120	3.9						
SCMS-GC-PD (ECD) (Liao 2001)	$CHCl_3$	0.67	93	3	YES / NO	Instrum. Anal. / Med-High	15-18K / YES	28-30 min / 2-2.1 per hr	Readily Adaptable / Near Real Time	Injection valve, Fast loop-Flow Cell Arrangement
	$CHBrCl_2$	0.30	101	1						
	$CHBr_2Cl$	1.47	100	5						
	$CHBr_3$	2.23	120	1						
	TTHMs	4.67	104	3						
Direct Aqueous injection-GC-ECD (Biziulk et al. 1996)	$CHCl_3$	0.01	NR	1.74	YES / NO	Instrum. Anal. / Low-Med	12-14K / YES**	NR	Possibly Adaptable / Near Real Time	Automated GC injector, FIA or SILLEX techniques
	$CHBrCl_2$	NR								
	$CHBr_2Cl$	0.01								
	$CHBr_3$	NR								
	TTHMs	0.02								
Liquid-Liquid Extraction GC-MS (Nikolaou et al. 2002b)	$CHCl_3$	0.01	115	11	YES / NO	Instrum. Anal. / Med-High	65-70K / YES§	65-75 / 0.8-0.9	Possibly Adaptable / Near Real Time	Automated sampler, FIA or SILLEX techniques
	$CHBrCl_2$	0.02	115	12						
	$CHBr_2Cl$	0.02	109	9						
	$CHBr_3$	0.03	112	13						
	TTHMs	0.08	113	11						

(continued)

Table 4.1 (Continued)

Analytical method (Reference)*	THM species	MDL (µg/L)	Mean % recovery	% RSD	Comm. avail. / EPA basis	Equipt. required / Est. skill level†	Instru. cost ($)‡ / Rugged and portable	Anal. time / Samp rate	Adapt to on-line monitoring / Real or near real time	Possible automation modes
Purge and Trap-GC-MS (Gelover et al. 2000)	$CHCl_3$ $CHBrCl_2$ $CHBr_2Cl$ $CHBr_3$ TTHMs	NR	NR	NR	YES / YES	Instrum. Anal. / Med-High	68-70K / YES§	50 min / 1.2 per hr	Readily Adaptable / Near Real Time	Continuous flow Purge and Trap; SCMS-GC, SPME, HS-SPME
HS-GC-MS (Takahashi et al. 2003)	$CHCl_3$ $CHBrCl_2$ $CHBr_2Cl$ $CHBr_3$ TTHMs	NR	NR	NR	YES / YES	Instrum. Anal. / Med-High	65-68K / YES§	75-80 min / 0.8 per hr	Possibly Adaptable / Near Real Time	Dynamic Headspace sampling
HS Sampling-GC-ECD (Batterman et al. 2002)	$CHCl_3$ $CHBrCl_2$ $CHBr_2Cl$ $CHBr_3$ TTHMs	0.1 0.03 0.04 0.5 0.67	NR	NR	YES / YES	Instrum. Anal. / Medium	20-22K / YES**	105 min / 0.6 per hr	Possibly Adaptable / Near Real Time	Dynamic Headspace sampling
HS-GC-MS (Nikolaou et al. 2002b)	$CHCl_3$ $CHBrCl_2$ $CHBr_2Cl$ $CHBr_3$ TTHMs	0.20 0.05 0.05 0.10 0.22	88 98 98 99 96	17 14 12 15 15	YES / YES	Instrum. Anal. / Med-High	65-68K / YES§	70-85 min / 0.7-0.9 per hr	Possibly Adaptable / Near Real-Time	Dynamic Headspace sampling

(continued)

Table 4.1 (Continued)

Analytical method (Reference)*	THM species	MDL (µg/L)	Mean % recovery	% RSD	Comm. avail. / EPA basis	Equipt. required / Est. skill level†	Instru. cost ($)‡ / Rugged and portable	Anal. time / Samp rate	Adapt to on-line monitoring / Real or near real time	Possible automation modes
Automated MIMS (Lopez-Avila et al. 1999)	$CHCl_3$	At µg/L level	NR	5.7	YES / NO	Instrum. Anal. / High	63-65K / YES§	5-6 min / 10-12 per hr	Readily Adaptable / Near Real Time	FIA / SIA techniques coupled to MIMS
	$CHBrCl_2$			NR						
	$CHBr_2Cl$			NR						
	$CHBr_3$			NR						
	TTHMs			NR						
Membrane Introduction-Fast-GC (Chang and Her 2000)	$CHCl_3$	0.002	92-97%	1.4-4.4	Combo of comm. and non-comm. / NO	Instrum. Anal. / High	72-74K / YES§	4 min / ~15 per hr	Readily Adaptable / Near Real Time	FIA / SIA techniques coupled to MIMS
	$CHBrCl_2$	0.004								
	$CHBr_2Cl$	0.004								
	$CHBr_3$	0.008								
	TTHMs	0.018								
SPME-GC-MS (Wang and Tan 2000)	$CHCl_3$	0.04	98-100%	2.5-6.2	YES / YES	Instrum. Anal. / Med-High	64-67K / YES§	20 min / 3 per hr	Possibly Adaptable / Near Real Time	Automated sampler for GC, FIA / SIA possibilities
	$CHBrCl_2$	0.04								
	$CHBr_2Cl$	0.08								
	$CHBr_3$	0.09								
	TTHMs	0.25								
HS-SPME GC-MS (Stack et al. 2000)	$CHCl_3$	2.8	Varied by up to 20%	0.9-19	YES / YES	Instrum. Anal. / Med-High	64-67K / YES§	30 min / 2 per hr	Possibly Adaptable / Near Real Time	Automated sampler for GC, FIA / SIA possibilities
	$CHBrCl_2$	1.4								
	$CHBr_2Cl$	1.0								
	$CHBr_3$	1.2								
	TTHMs	6.4								

(continued)

Table 4.1 (Continued)

Analytical method (Reference)*	THM species	MDL (μg/L)	Mean % recovery	% RSD	Comm. avail. / EPA basis	Equipt. required / Est. skill level†	Instru. cost ($)‡ / Rugged and portable	Anal. time / Samp rate	Adapt to on-line monitoring / Real or near real time	Possible automation modes
SPME-GC-MS (Guidotti and Ravajoli 1999)	$CHCl_3$	0.125	Varied by 20%	7.8-13.2	YES / YES	Instrum. Anal. / Med-High	63-65K / YES§	50 min / 1.2 per hr	Possibly Adaptable / Near Real Time	Automated sampler for GC, FIA / SIA possibilities
	$CHBrCl_2$	0.15								
	$CHBr_2Cl$	0.09								
	$CHBr_3$	0.105								
	TTHMs	0.47								
Liquid-liquid Extraction-GC-MS (Brush and Rice 1994)	$CHCl_3$	NR	NR	20-50	YES / YES	Instrum. Anal. / Med-High	60-62K / YES§	75 min / 0.8 per hr	Possibly adaptable / Near Real Time	FIA or SILLEX techniques
	$CHBrCl_2$									
	$CHBr_2Cl$									
	$CHBr_3$									
	TTHMs									
Colorimetric Method (Reckhow and Pierce 1992)	$CHCl_3$	8 – 12	NR	15-22	YES / NO	UV-VIS-Wet Chem / Low-Med	2-10K / YES	15-30 min / 2-3 per hr	Readily Adaptable / Near Real Time	FIA / SIA, possibilities; Limit to total conc.; No monohalo. Species detected
	$CHBrCl_2$									
	$CHBr_2Cl$									
	$CHBr_3$									
	TTHMs									
Continuous flow injection-fluorescent (Aoki and Kawakami 1989)	$CHCl_3$	0.8	84-94	4.4	YES / NO	FIA-Instrum. Anal. / Low-Med	<15K / YES	15-30 min / 2-3 per hr	Readily Adaptable / Near Real Time	FIA / SIA, possibilities; Limit to total conc.; No monohalo. Species detected
	$CHBrCl_2$	NR		NR						
	$CHBr_2Cl$	NR		NR						
	$CHBr_3$	NR		NR						
	TTHMs	NR		NR						

(continued)

Table 4.1 (Continued)

Analytical method (Reference)*	THM species	MDL (μg/L)	Mean % recovery	% RSD	Comm. avail. / EPA basis	Equipt. required / Est. skill level†	Instru. cost ($)‡ / Rugged and portable	Anal. time / Samp rate	Adapt to on-line monitoring / Real or near real time	Possible automation modes
A rapid colorimetric method (McKenzie, Studabaker and Carter 1997)	$CHCl_3$ $CHBrCl_2$ $CHBr_2Cl$ $CHBr_3$ TTHMs	5	74-100	10	YES / NO	UV-VIS-Wet Chem / Low	2-10K / YES	15-30 min / 2-3 per hr	Readily Adaptable / Near Real Time	FIA / SIA, possibilities; Limit to total conc.; No monohalo. Species detected.
A simple colorimetric method (Lord 1999)	$CHCl_3$ $CHBrCl_2$ $CHBr_2Cl$ $CHBr_3$ TTHMs	6	107	5.8	YES / NO	UV-VIS-Wet Chem / Low	2-10K / YES	30 min / 2 per hr	Readily Adaptable / Near Real Time	FIA / SIA, possibilities; Limit to total conc.; No monohalo. Species detected.

* Due to space limitations, the abbreviated reference in the table lists only the first author of the cited paper followed by the publication year.

† The skill level refers to the level of ability required by the analyst. The rating infers that traditional wet chemical/colorimetric chemistries require a generally low level of training, common chromatographic based methods (i.e., non-mass spectrometric methods) are rated medium, and specialized techniques including mass spectrometric methods require a high level of analyst skill.

‡ Instrument cost will certainly vary from vendor to vendor. The estimated instrument costs are based on "low-end" instrumentation and specialized materials designed specifically to fit the needs of the particular analysis.

§ There are portable GC-MS instruments that are commercially available. They are generally considered too costly to meet the terms of the Ideal Method. The estimated prices for GC-MS instruments presented in the paper are for typical "bench-top" models.

** The use of an ECD is generally restricted by a federal/state license. It is not generally accepted practice to move them from location to location as would be necessary for a portable instrument. In addition, there are issues related to venting the ECD exhaust in some instances.

NR "Not Reported" – meaning that the specified information was not reported in the paper or otherwise not available.

Table 4.2
Summary of analytical methods from the literature for measuring the concentrations of five HAAs in drinking water

Analytical method (Reference)*	THM species	MDL (µg/L)	Mean % recovery	% RSD	Comm. avail. / EPA basis	Equipt. required / Est. skill level†	Instru. cost ($)‡ / Rugged and portable	Anal. time / Samp rate	Adapt to on-line monitoring / Real or near real time	Possible automation modes
USEPA 552 LLE-Derivatization, GC-ECD (USEPA 1990b)	MCAA	0.052	74	38	YES / YES	Instrum. Anal. / Medium	14-16K / YES**	150 min / 0.4 per hr	Possibly Adaptable / Near Real Time	FIA, SIA or SILEX techniques
	MBAA	0.074	58	7						
	DCAA	0.015	84	10						
	TCAA	0.085	78	59						
	DBAA	0.015	58	14						
USEPA 552.1 LLE-Derivatization, GC-ECD (USEPA 1992)	MCAA	0.21	98	4.6	YES / YES	Instrum. Anal. / Medium	14-16K / YES**	320 min / 0.2 per hr	Possibly Adaptable / Near Real Time	FIA, SIA or SILEX techniques
	MBAA	0.24	73	7.9						
	DCAA	0.45	110	7.7						
	TCAA	0.07	60	4.0						
	DBAA	0.09	58	6.4						
USEPA 552.2 LLE, Derivatization, GC-ECD (USEPA 1995d)	MCAA	0.273	94.7	7.3	YES / YES	Instrum. Anal. / Medium	14-16K / YES**	290 min / 0.2 per hr	Possibly Adaptable / Near Real Time	FIA, SIA or SILEX techniques
	MBAA	0.204	102	5.0						
	DCAA	0.242	84.7	9.6						
	TCAA	0.079	93.0	10						
	DBAA	0.066	95.4	9.2						
USEPA 552.3 Micro-LLE, Derivatization, GC-ECD (MTBE) (USEPA 2003b)	MCAA	0.17	95.8	4.0	YES / YES	Instrum. Anal. / Medium	14-16K / YES**	~160 min / 0.4 per hr	Possibly Adaptable / Near Real Time	FIA, SIA or SILEX techniques
	MBAA	0.027	92.2	1.8						
	DCAA	0.020	93.8	1.6						
	TCAA	0.019	105	0.52						
	DBAA	0.012	105	0.63						

(continued)

Table 4.2 (Continued)

Analytical method (Reference)*	THM species	MDL (µg/L)	Mean % recovery	% RSD	Comm. avail. / EPA basis	Equipt. required / Est. skill level†	Instru. cost ($)‡ / Rugged and portable	Anal. time / Samp rate	Adapt to on-line monitoring / Real or near real time	Possible automation modes
LLE, derivatization (methanolic), GC-ECD (Nikolaou et al. 2002b)	MCAA MBAA DCAA TCAA DBAA	0.20 0.05 0.02 0.01 0.02	72.8 99.9 96.8 102 107	1.2 7 5.2 4.8 5.6	YES / YES	Instrum. Anal. / Medium	14-16K / YES**	140 min / 0.4 per hr	Possibly Adaptable / Near Real Time	FIA, SIA, SILEX or LOV techniques
LLE, derivatization (methanolic), GC-ECD (Humbert and Fernandez 1976)	TCAA	NR	NR	5.3	YES / NO	Instrum. Anal. / Medium	14-16K / YES**	100 min / 0.4 per hr	Possible, but not desirable / Near Real Time	Not desirable to automate, older packed column technology
LLE, esterification, GC-ECD (Vesterberg, Gorczak and Krasts 1975)	TCAA	NR	NR	NR	YES / YES	Instrum. Anal. / Medium	14-16K / YES**	1230 min / 0.04 per hr	Possible, but not desirable / Near Real Time	Not desirable to automate, older packed column technology
LLE, derivatization (methanolic), GC-ECD (Donnell et al. 1995)	TCAA	0.05	99.6	3.5	YES / YES	Instrum. Anal. / Medium	14-16K YES**	130 min / 0.4 per hr	Possibly Adaptable / Near Real Time	FIA, SIA, SILEX or LOV techniques
Acidic methanol esterification, GC-ECD†† (Cancho, Ventura and Galceran 1999)	MCAA DCAA TCAA MBAA DBAA	NR	NR	NR	YES / YES	Instrum. Anal. / Medium	14-16K / YES**	140 min / 0.4 per min	Possibly Adaptable / Near Real Time	FIA, SIA, SILEX or LOV techniques

(continued)

Table 4.2 (Continued)

Analytical method (Reference)*	THM species	MDL (µg/L)	Mean % recovery	% RSD	Comm. avail. / EPA basis	Equipt. required / Est. skill level†	Instru. cost ($)‡ / Rugged and portable	Anal. time / Samp rate	Adapt to on-line monitoring / Real or near real time	Possible automation modes
Micro-LLE, SPME-GC-ECD†† (Wu, Gabryelski and Froese 2002)	MCAA DCAA TCAA MBAA DBAA	0.6-1.2	NR	NR	YES / YES	Instrum. Anal. / Med-High	17-19K / YES**	140 min / 0.4 per hr	Possibly Adaptable / Near Real Time	Autosampler, FIA, SIA, SILEX, LOV or microfluidic based techniques
Simultaneous extraction, GC-ECD (Benanou, Acobas and Sztajnbok 1998)	MCAA DCAA TCAA MBAA DBAA	0.1-0.2	NR	10-20	YES / NO	Instrum. Anal. / Med-High	14-16K / YES**	130 min / 0.4 per hr	Possibly Adaptable / Near Real Time	FIA, SIA, SILEX or LOV techniques
Large volume injection GC-ECD (Drechsel et al. 2001)	TCAA	0.1	NR	1-9	YES / NO	Instrum. Anal. / Med-High	12-14K / YES**	45 min / 1.3 per hr	Possibly Adaptable / Near Real Time	FIA, SIA, SILEX or LOV techniques
Micro-LLE, derivatization (methanolic), GC-MS†† (Xie 2001)	MCAA MBAA DCAA TCAA DBAA	0.21 0.71 0.07 0.15 0.04	87 73 92 137 93	NR	YES / YES	Instrum. Anal. / High	60-62K / YES§	110 min / 0.5 per hr	Possibly Adaptable / Near Real Time	FIA, SIA, SILEX or LOV techniques
Difluoroanilide derivatization, GC-MS (SIM) (Ozawa 1993)	MCAA MBAA DCAA TCAA DBAA	0.5 2 0.5 0.5 1	95 87 99 96 104	1.5 6.6 0.7 1.9 1.8	YES / NO	Instrum. Anal. / High	60-62K / YES§	60-70 min / 0.8 per hr	Possibly Adaptable / Near Real Time	FIA, SIA, SILEX or LOV techniques

(continued)

Table 4.2 (Continued)

Analytical method (Reference)*	THM species	MDL (µg/L)	Mean % recovery	% RSD	Comm. avail. / EPA basis	Equipt. required / Est. skill level†	Instru. cost ($)‡ / Rugged and portable	Anal. time / Samp rate	Adapt to on-line monitoring / Real or near real time	Possible automation modes
LLE-derivatization GC-MS (Kristiansen et al. 1996)	DCAA	0.2	69	9.4	YES / YES	Instrum. Anal. / Med-High	65-67K / YES§, **	100 min / 0.6 per hr	Possibly Adaptable / Near Real Time	FIA, SIA, SILEX or LOV techniques
	TCAA	0.1	70	9.5						
	DBAA	0.15	84	21						
SPE-GC-MS derivatization (diazomethane) (Martinez 1998b)	MCAA	20	18	NR	YES / YES	Instrum. Anal. / High	62-65K / YES§	90 min / 0.7 per hr	Possibly Adaptable / Near Real Time	Autosampler, FIA, SIA, SILEX, LOV or microfluidic based techniques
	MBAA	20	22	NR						
	DCAA	3	35	12						
	TCAA	5	42	NR						
	DBAA	20	40	4						
Micro-LLE, acidic MeOH methylation GC-ECD (Xie, Reckhow and Springborg 1998)	DCAA	0.2	111	NR	YES / YES	Instrum. Anal. / Medium	12-14K / YES **	130 min / 0.4 per hr	Possibly Adaptable / Near Real Time	FIA, SIA, SILEX, or LOV based techniques
	TCAA	0.09	109							
	MBAA	0.4	118							
	DBAA	0.07	108							
LLE, derivatization (diazomethane) GC-MS, GC-ECD method (Clemens and Schoeler 1992)	MCAA	1.5	82	NR	YES / YES	Instrum. Anal. / Med-High	62-67K / YES§, **	190 min / 0.3 per hr	Possibly Adaptable / Near Real Time	FIA, SIA, SILEX, or LOV based techniques
	DCAA	0.12	92							
	TCAA	0.05	79							
In situ anilide derivatization GC/MS†† (Scott and Alaee 1998)	MCAA, MBAA, DCAA, TCAA, BCAA	1	NR	10-41	YES / NO	Instrum. Anal. / High	75-80K / NO	135 min / 0.4 per hr	Possibly Adaptable / Near Real Time	FIA, SIA, SILEX, or LOV based techniques

(continued)

Table 4.2 (Continued)

Analytical method (Reference)*	THM species	MDL (µg/L)	Mean % recovery	% RSD	Comm. avail. / EPA basis	Equipt. required / Est. skill level†	Instru. cost ($)‡ / Rugged and portable	Anal. time / Samp rate	Adapt to on-line monitoring / Real or near real time	Possible automation modes
Acidic ethanolic derivatization, SPME, GC-ITS-MS (Sarrion, Santos and Galceran 1999)	MCAA	0.20	NR	7.2	YES / NO	Instrum. Anal. / High	63-65K / YES§	90 min / 0.7 per hr	Possibly Adaptable / Near Real Time	Autosampler, FIA, SIA, SILEX or LOV techniques
	MBAA	0.04		7.7						
	DCAA	0.01		6.3						
	TCAA	0.02		6.7						
	DBAA	0.03		7.2						
Acidic diethyl sulfate derivatization SPME , GC-ITS-MS (Sarrion, Santos and Galceran 2000)	MCAA	0.20	NR	10.9	YES / NO	Instrum. Anal. / High	62-65K / YES§	80 min / 0.8 per hr	Possibly Adaptable / Near Real Time	Autosampler FIA, SIA, SILEX or LOV techniques
	MBAA	0.40		9.8						
	DCAA	0.07		8.0						
	TCAA	0.02		9.0						
	DBAA	0.01		7.2						
Lichrolut-EN-non-suppressed IC (Sarzanini, Bryzzoniti and Mentasti 1999)	MCAA	NR	NR	NR	YES / NO	Instrum. Anal. / Medium	17-25K / YES	50 min / 1.2 per hr	Readily Adaptable / Near Real Time	Automated injection valve, Post-column sensitivity enhancement
	MBAA	20	80.3	8.4						
	DCAA	7.0	83.7	13						
	TCAA	10	81.8	11						
	DBAA	10	73.1	13						
Activated carbon-suppressed IC (Sarzanini, Bryzzoniti and Mentasti 1999)	MCAA	NR	NR	NR	YES / NO	Instrum. Anal. / Medium	17-25K / YES	50 min / 1.2 per hr	Readily Adaptable / Near Real Time	Automated injection valve, Post-column sensitivity enhancement
	MBAA		5.4	0.9						
	DCAA		22.9	8.2						
	TCAA		55.7	16						
	DBAA		25.8	10						

(continued)

Table 4.2 (Continued)

Analytical method (Reference)*	THM species	MDL (µg/L)	Mean % recovery	% RSD	Comm. avail. / EPA basis	Equipt. required / Est. skill level†	Instru. cost ($)‡ / Rugged and portable	Anal. time / Samp rate	Adapt to on-line monitoring / Real or near real time	Possible automation modes
IC, anion exchange with suppressed cond. detection (Nair, Saari-Nordhaus and Anderson 1994)	MCAA	8	NR	NR	YES / NO	Instrum. Anal. / Medium	18-25K / YES	40 min / 1.5 per hr	Readily Adaptable / Near Real Time	Automated injection valve, Post-column sensitivity enhancement
	MBAA	85								
	DCAA	16								
	TCAA	80								
	DBAA	30								
IC, anion exclusion with UV (210nm) detection (Nair, Saari-Nordhaus and Anderson 1994)	MCAA	70	NR	NR	YES / NO	Instrum. Anal. / Medium	18-25K / YES	40 min / 1.5 per hr	Readily Adaptable / Near Real Time	Automated injection valve, Post-column sensitivity enhancement
	MBAA	85								
	DCAA	8								
	TCAA	5.1								
	DBAA	90								
IC with electrochemical detector with pyrolytic graphite (Carrero and Rusling 1999)	MCAA	10000	NR	NR	YES / NO	Instrum. Anal. / Med-High	18-20K / YES	50 min / 1.2 per hr	Not Desirable	Not desirable to automate
	MBAA	480								
	DCAA	4000								
	TCAA	120								
	DBAA	160								
HPLC, reversed-phase C-18 column, UV at 210 nm (Husain et al. 1992)	MCAA	NR	NR	NR	YES / NO	Instrum. Anal. / Medium	17-25K / YES	40 min / 1.5 per hr	Readily Adaptable / Near Real Time	Automated injection valve, Post-column sensitivity enhancement
	MBAA	NR								
	DCAA	NR								
	TCAA	10								
	DBAA	NR								

(continued)

Table 4.2 (Continued)

Analytical method (Reference)*	THM species	MDL (µg/L)	Mean % recovery	% RSD	Comm. avail. / EPA basis	Equipt. required / Est. skill level†	Instru. cost ($)‡ / Rugged and portable	Anal. time / Samp rate	Adapt to on-line monitoring / Real or near real time	Possible automation modes
Polymer based micro-electrodes (Akhtar, Too and Wallace 1997)	MCAA	1	NR	NR	NO / NO	Instrum. Anal. / High	NR	45 min / 1.3 per hr	Not Desirable	Not desirable to automate
	DCAA	10								
	TCAA	100								
	DBAA	100								
Reversed phase ion pairing HPLC (Vichot and Furton 1994)	MCAA	2000	NR	NR	YES / NO	Instrum. Anal. / Medium	17-25K / YES	40 min / 1.5 per hr	Readily Adaptable / Near Real Time	Not desirable to automate
	MBAA	8000								
	DCAA	10000								
	DBAA	15000								
Large Volume direct injection by IC†† (Liu et al. 2003)	MCAA	1.91	103	4.45	YES / NO	Instrum. Anal. / Medium	17-25K / YES	55 min / 1 per hr	Readily Adaptable / Near Real Time	Automated injection valve, Post-column sensitivity enhancement
	DCAA	4.62	93.5	3.11						
	TCAA	5.11	91.3	3.99						
	MBAA	3.11	92.2	5.66						
LC-ESI-MS (Hashimoto and Otsuki 1998)	MCAA	0.029	96	3	YES / NO	Instrum. Anal. / High	140-180K / NO	60 min / 1 per hr	Likely not a realistic choice for this project	FIA techniques, likely too expensive for this project
	MBAA	0.027	92	5						
	DCAA	0.003	92	6						
	TCAA	0.070	85	3						
	DBAA	0.014	93	5						
LC-ESI-MS†† (Takino, Daishima and Yamaguchi 2000)	MCAA	0.089	NR	4.5	YES / NO	Instrum. Anal. / High	140-180K / NO	60 min / 1 per hr	Likely not a realistic choice for this project	FIA techniques, likely too expensive for this project
	MBAA	0.096		5.3						
	DCAA	0.024		1.5						
	TCAA	0.083		4.2						
	DBAA	0.091		4.9						

(continued)

Table 4.2 (Continued)

Analytical method (Reference)*	THM species	MDL (µg/L)	Mean % recovery	% RSD	Comm. avail. / EPA basis	Equipt. required / Est. skill level†	Instru. cost ($)‡ / Rugged and portable	Anal. time / Samp rate	Adapt to on-line monitoring / Real or near real time	Possible automation modes
Liquid-liquid microextraction ESI-MS†† (Magnuson and Kelty 2000)	MCAA	0.60	NR	NR	YES / NO	Instrum. Anal. / High	140-180K / NO	60 min / 1 per hr	Likely not a realistic choice for this project	FIA or SILEX techniques, likely too expensive for this project
	DCAA	0.32								
	TCAA	0.13								
	MBAA	0.62								
	DBAA	0.33								
Solid phase extraction LC-ESI-MS†† (Loos and Barcelo 2001)	MCAA	1.6	26	3	YES / NO	Instrum. Anal. / High	140-180K / NO	40 min / 1.5 per hr	Likely not a realistic choice for this project	FIA techniques, likely too expensive for this project
	MBAA	0.8	42	6						
	DCAA	0.8	55	5						
	DBAA	1.3	56	7						
	TCAA	0.9	75	5						
Solid phase extraction HPLC-ESI-MS/MS (Kuklenyik, Ashley and Calafal 2002)	TCAA	0.5	97-124	8.5	YES / NO	Instrum. Anal. / High	200-250K / NO	30 min / 1 per hr	Possibly Adaptable / Near-Real Time	FIA techniques, likely too expensive for this project
On-line preconcentration IC-ESI-MS†† (Roehl et al. 2002)	MCAA	3	NR	NR	YES / NO	Instrum. Anal. / High	140-180K / NO	50 min / 1.2 per hr	Possibly Adaptable / Near-Real Time	FIA techniques, likely too expensive for this project
	MBAA	2								
	DCAA	3								
	DBAA	1								
	TCAA	1								

(continued)

Table 4.2 (Continued)

Analytical method (Reference)*	THM species	MDL (µg/L)	Mean % recovery	% RSD	Comm. avail. / EPA basis	Equipt. required / Est. skill level†	Instru. cost ($)‡ / Rugged and portable	Anal. time / Samp rate	Adapt to on-line monitoring / Real or near real time	Possible automation modes
CE and CZE (Martinez 1999b)	MCAA	5	NR	9.8	YES / NO	Instrum. Anal. / Med-High	25-30K / NO	45 min / 1.3 per hr	Possibly Adaptable / Near-Real Time	LOV, microfluidic device FIA/SIA
	MBAA	5		6.3						
	DCAA	2		5.5						
	DBAA	3		7.4						
	TCAA	2		6.7						
LLE- CZE (Kim et al. 2001)	MCAA	5	90.2	10.3	YES / NO	Instrum. Anal. / Med-High	25-30K / NO	45 min / 1.3 per hr	Possibly Adaptable / Near-Real Time	SILEX, LOV, microfluidic device FIA/SIA
	MBAA	5	88.5	6.9						
	DCAA	5	95.8	9.5						
	DBAA	5	90.4	6.1						
CZE (Martinez 1999a)	MCAA	0.4	54	7.3	YES / NO	Instrum. Anal. / Med-High	25-30K / NO	45 min / 1.3 per hr	Possibly Adaptable / Near-Real Time	LOV, FIA, SIA or microfluidic device
	MBAA	NR	83	8.3						
	DCAA	NR	60	6.6						
	TCAA	NR	58	10.2						
	DBAA	0.9	58	14.5						
CZE (Martinez 1998a)	MCAA	0.40	66	1.1	YES / NO	Instrum. Anal. / Med-High	25-30K / NO	45 min / 1.3 per hr	Possibly Adaptable / Near-Real Time	LOV, FIA, SIA or microfluidic device
	MBAA	0.75	60	2.0						
	DCAA	0.15	70	1.5						
	TCAA	0.50	80	1.9						
	DBAA	0.90	70	3.4						

(continued)

Table 4.2 (Continued)

Analytical method (Reference)*	THM species	MDL (µg/L)	Mean % recovery	% RSD	Comm. avail. / EPA basis	Equipt. required / Est. skill level†	Instru. cost ($)‡ / Rugged and portable	Anal. time / Samp rate	Adapt to on-line monitoring / Real or near real time	Possible automation modes
CE-API-MS†† (Ahrer and Buchberger 1999)	MCAA	1.7	33	15.2	YES / NO	Instrum. Anal. / High	200-250K / NO	45 min / 1.3 per hr	Possibly Adaptable / Near-Real Time	FIA, SIA , microfluidic techniques LOV
	MBAA	1.3	36	7.8						
	DCAA	0.3	72	8.2						
	TCAA	0.5	69	4.5						
	DBAA	1.0	94	5.1						
CE†† (Xie, Zhou and Romano 2000)	MCAA	0.21	87	NR	YES / NO	Instrum. Anal. / Med-High	25-30K / NO	110 min / 0.5 per min	Possibly Adaptable / Near-Real Time	LOV, FIA, SIA or microfluidic device
	DCAA	0.07	92							
	TCAA	0.15	137							
CE†† (Soga and Imaizumi 2001)	MBAA	0.71	73	NR	YES / NO	Instrum. Anal. / Med-High	20-25K / NO	110 min / 0.5 per hr	Possibly Adaptable / Near-Real Time	LOV, FIA, SIA or microfluidic device
	DBAA	0.04	93							
Colorimetric Method (Ogata, Takatsuka and Tomokuni 1970)	MCAA	NR	NR	NR	YES / NO	UV-Vis Wet Chemical / Low	2-15K / YES	20 min / 3 per hr	Readily Adaptable / Near Real Time	FIA / SIA, possibilities; Limit to total conc.; No monohalo. Species detected
	MBAA		NR							
	DCAA		NR							
	TCAA		95							
	DBAA		NR							

(continued)

Table 4.2 (Continued)

Analytical method (Reference)*	THM species	MDL (μg/L)	Mean % recovery	% RSD	Comm. avail. / EPA basis	Equipt. required / Est. skill level†	Instru. cost ($)‡ / Rugged and portable	Anal. time / Samp rate	Adapt to on-line monitoring / Real or near real time	Possible automation modes
Colorimetric Method (Tanaka and Ikeda 1968)	MCAA MBAA DCAA TCAA DBAA	NR	NR NR NR 96.9 NR	NR NR NR 2.2 NR	YES / NO	UV-Vis Wet Chemical / Low	2-15K / YES	20 min / 3 per hr	Readily Adaptable / Near Real Time	FIA / SIA, possibilities; Limit to total conc.; No monohalo. Species detected
Colorimetric Method (Pillai, Rastogi and Gupta 1999)	MCAA MBAA DCAA TCAA DBAA	NR	NR NR NR 99 NR	NR	YES / NO	UV-Vis Wet Chemical / Low	2-15K / YES	20 min / 3 per hr	Readily Adaptable / Near Real Time	FIA / SIA, possibilities; Limit to total conc.; No monohalo. Species detected
Colorimetric Method (Pal 1996)	TCAA	NR	NR	7	YES / NO	UV-Vis Wet Chemical / Low	2-15K / YES	40 min / 1.5 per hr	Readily Adaptable / Near Real Time	FIA / SIA, possibilities; Limit to total conc.; No monohalo. Species detected
Colorimetric Method (Bhattacharjee, Cherian and Gupta 1991)	TCAA	NR	NR	3.4	YES / NO	UV-Vis Wet Chemical / Low	2-15K / YES	15 min / 4 per hr	Readily Adaptable / Near Real Time	FIA / SIA, possibilities; Limit to total conc.; No monohalo. Species detected
Fluorescence Spot Test (Rathore, Saxena and Begum 1992)	TCAA	NR	NR	NR	YES / NO	Wet Chemical / Low	None / YES	50 min / 1.2 per hr	Not Desirable	Not Desirable

(continued)

Table 4.2 (Continued)

Analytical method (Reference)*	THM species	MDL (µg/L)	Mean % recovery	% RSD	Comm. avail. / EPA basis	Equipt. required / Est. skill level†	Instru. cost ($)‡ / Rugged and portable	Anal. time / Samp rate	Adapt to on-line monitoring / Real or near real time	Possible automation modes
ESI-FAIMS-MS (Ells et al. 1999)	MCAA	0.7	NR	NR	NO / NO	NR	150-200K / NO	NR / NR	Likely not a realistic choice for this project	FIA
	MBAA	0.8								
	DCAA	1.4								
	TCAA	4								
	DBAA	0.5								
ESI-FAIMS-MS (Ells et al. 2000)	MCAA	0.028	NR	NR	NO / NO	NR	150-200K / NO	NR / NR	Likely not a realistic choice for this project	FIA
	MBAA	0.018								
	DCAA	0.010								
	TCAA	0.036								
	DBAA	0.014								
ESI-FAIMS-MS (Gabryelski, Wu and Froese 2003)	MCAA	0.31	NR	NR	NO / NO	NR	150-200K / NO	NR / NR	Likely not a realistic choice for this project	FIA
	MBAA	0.12								
	DCAA	0.17								
	TCAA	0.18								
	DBAA	0.07								

* Due to space limitations, the abbreviated reference in the table lists only the first author of the cited paper followed by the publication year.

† The skill level refers to the level of ability required by the analyst. The rating infers that traditional wet chemical/colorimetric chemistries require a generally low level of training, common chromatographic based methods (i.e., non-mass spectrometric methods) are rated medium, and specialized techniques including mass spectrometric methods require a high level of analyst skill.

‡ Instrument cost will certainly vary from vendor to vendor. The estimated instrument costs are based on "low-end" instrumentation and specialized materials designed specifically to fit the needs of the particular analysis.

§ There are portable GC-MS instruments that are commercially available. They are generally considered too costly to meet the terms of the Ideal Method. The estimated prices for GC-MS instruments presented in the paper are for typical "bench-top" models.

** The use of an ECD is generally restricted by a federal/state license. It is not generally accepted practice to move them from location to location as would be necessary for a portable instrument. In addition, there are issues related to venting the ECD exhaust in some instances.

†† These methods have been used for all nine HAAs.

NR stands for "Not Reported" – meaning that the specified information was not reported in the paper or otherwise not available.

Table 4.3
The GC columns used for separation of THMs

Column type	Description
DB-624	30 m × 0.53 mm i.d. × 3 μm f.t. (J&W Scientific, Inc.)
	75 m × 0.53 mm i.d. × 3 μm f.t. fused silica capillary (J&W Scientific, Inc.)
	25 m × 0.2 mm i.d. × 1.12 μm f.t.
	30 m × 0.32 mm × 1.8 μm f.t.
DB-5	30 m × 0.32 mm i.d. × 1 μm f.t. (J&W Scientific, Inc.)
	300.25 mm i.d. × 0.25 μm f.t. (J&W Scientific, Inc.)
	30 m × 0.32 mm i.d. fused silica capillary column
DB-5MS	5 m × 0.25 mm i.d × 0.25 μm f.t. (J&W scientific)
DB-1	30 m × 0.32 mm i.d. fused silica capillary with chemically bonded methyl polysiloxane phase
	30 m × 0.25 mm i.d. fused silica capillary with chemically bonded methyl polysiloxane phase
	30 m × 0.32 mm i.d. fused-silica capillary column, coated with bonded 5μm apolar DB-1 phase
	30m × 0.32 mm i.d. × 0.25 μm f.t.
DB-210	30 m × 0.32 mm i.d. with chemically bonded 50% trifluoropropyl phase
DB-1301	60 m × 0.25 mm i.d. × 0.25 μm f.t.
VB-1	30 m × 0.32 mm i.d. × 1.0 μm f.t.
HP-5MS	30 m × 0.25 mm i.d. × 1.0 μm f.t.
	30 m × 0.25 mm i.d. × 0.25 μm f.t.
HP-1	50 m × 0.2 mm i.d. × 0.5 μm f.t.
	12 m × 0.2 3 mm i.d. crosslinked methyl silicone column
HP-VOC	60 m × 0.32 mm i.d. × 1.8 μm fused silica capillary
Others	60 m × 0.75 mm i.d. VOCOL (Supelco, Inc.) wide-bore capillary column with 1.5 μm f.t.
	105 m × 0.53 mm i.d., RTX-502.2 (O.I Corporation/RESTEK Corporation) mega-bore capillary column, with 3.0 μm f.t., or equivalent
	30 m × 0.25 mm i.d. with chemically bonded 6% cyanopropylphenyl/94% dimethyl polysiloxane phase
	Chrompack CP-SIL 13 CB, 25 m × 0.32 mm i.d. f.t. 1.2 μm
	6 ft × 1/8 in. aluminum column packed with 3% SP-1000
	6 ft × 1/8 in. stainless steel columns (1% SP-1000 on 60/80 Carbopack B)
	60 m × 0.32 mm i.d. × 1.8 μm f.t.
	30 m × 0.22 mm × 0.3 μm 5% phenylmethyl silicone capillary column
	2.0 m × 4 mm i.d stainless steel column packed with 20% THEED on 60/90 mesh firebrick
	5 m × 0.25 mm i.d. × 0.25 μm f.t.

Table 4.4
The GC columns used for separation of HAAs

Column type	Description
DB-5	30 m × 0.25 mm i.d. × 0.25 μm f.t.
	15 m × 0.53 mm i.d. × .5 μm f.t.
DB-5MS	30 m × 0.25 mm i.d. × 25 μm f.t.
DB-1701	30 m × 0.32 mm i.d. × 0.25 μm f.t.
	30 m × 0.25 mm i.d. × 0.25 μm f.t.
DB-210	30 m × 0.32 mm i.d., 0.50 μm f.t.
DB-17	30 m × 0.32 mm i.d. × 0.25 μm f.t.
	15 m × 0.53 mm i.d. × 1.0 μm f.t.
	30 m × 0.25 mm i.d. × 0.25 μm f.t.
DB-5.625	30 m × 0.25 mm i.d. × 0.25 μm f.t.
	30 m × 0.32 mm i.d. × 0.25 μm f.t.
DB-1	30 m × 0.32 mm i.d. × 0.25 μm f.t.
	30 m × 0.25 mm i.d. × 1.0 μm f.t.
HP-5 MS	30 m × 0.25 mm i.d. × 0.25 μm f.t.
Others	2.0 m × 3.2 mm glass tube containing 5% OV – 17 on 100 – 120 mesh Gaschrom O
	Stainless steel with inner diameter of 1.9 mm, 210 cm long, and was packed with 15% SE 30 on Chromosorb W, A.W. 100 -120 m
	30 m × 0.32 mm i.d.,Varian-Chrompack
	25 m × 0.32 mm i.d.1 × .2 μm f.t., CO-Sil 8 CB

Table 4.5
Internal standards and surrogate analytes

	Internal Standards
THMs	Bromochloromethane
	2-bromo-1-chloropropane
	1,4-dichlorobutane
	toluene-d8
	1,1,2-trichloroethane
HAAs	Bromochloromethane
	2-bromo-1-chloropropane
	1,3-dibromobutane
	1,2-dibromopropane
	1,3-dibromopropane
	2,3-dichloropropionic acid
	ether solution of bromochloromethane +1,2-dibromopropane
	[$^2H_8$] naphthalene solution
	1,2,3-trichloropropane
	trifluoroacetic acid

	Surrogate Analytes
HAAs	2-bromopropionic acid
	2,3-dibromopropionic acid
	3,5-dichlorobenzoic acid
	2,3-dichloropropanoic acid

Table 4.6
The LC columns used for separation of HAAs

Column type	Description
C_{18} 100 RP-18	10 µm LiChrospher (250 × 4 mm i.d.)
Ion Pac AS9, AS10, AS11 and AS10	250 × 4 mm i.d., 250 × 2 mm i.d.
Alltech universal anion 300 column	150 × 4.6 mm
Wescan Anion Exclusion Column	300 × 7.8 mm
IonPac AS9-HC and AS16	250 × 2 mm i.d, 250 × 4 mm i.d.
Supelcogel C-610H HPLC column	7.8 mm i.d. × 30 cm long
RAM alkyl – diol silica (ADS) column	Packed with LiChrospher® RP – 18 ADS 25 µm, silica pore diameter 60 A (Merck, Darmstadt, Germany)
ODS3	150 × 2.1 mm i.d. with Inertsil ODS3 (GL Science, Tokyo)
ODS3	150 × 2.1 mm i.d. column packed with 5 µm
RP 2	Prism RP 2 × 100 mm, 3 µm
Fused silica capillary	75 µm inside diameter × 365 µm outside diameter × 93 cm from Supelco
Waters Accu – Sep polyamide – coated fused – silica capillary	(60 cm × 75 µm i.d.)
Fused silica capillaries	50 µm i.d. × 112.5 cm length

REFERENCES

Ahrer, W., and W. Buchberger. 1999. Determination of Haloacetic Acids by the Combination of Non-aqueous Capillary Electrophoresis and Mass Spectrometry. *Fresenius J. Anal. Chem.*, 365(7): 604-609.

Akhtar, P., C. O. Too, and G. G. Wallace. 1997. Detection of Haloacetic Acids at Conductive Electroactive Polymer-Modified Micro Electrodes. *Anal. Chim. Acta*, 341(2-3): 141-153.

Allonier, A.-S., M. Khalanski, A. Bermond, and V. Camel. 2000. Determination of Trihalomethanes in Chlorinated Sea Water Samples Using a Purge-and-Trap System Coupled to Gas Chromatograph. *Talanta*, 51(3): 467-477.

Aoki, T., and K. Kawakami. 1989. Continuous Flow Method for Determination of Total Trihalomethanes in Drinking Water With Membrane Separation. *Wat. Res.*, 23(6): 739-742.

Batterman, S., L. Zhang, S. Wang, and A. Franzblau. 2002. Partition Coefficients for the Trihalomethanes Among Blood, Urine, Water, Milk and Air. *Sci. Total Environ.*, 284(1-3): 237-247.

Benanou, D., F. Acobas, and P. Sztajnbok. 1998. Analysis of Haloacetic Acids in Water by a Novel Technique: Simultaneous Extraction—Derivatization. *Wat. Res.*, 32(9): 2798-2806.

Bhattacharjee, M., L. Cherian, and V. K. Gupta. 1991. Modified Fujiwara Reaction for the Determination of Trichloroacetic Acid. *Microchem. J.*, 43(2): 109-111.

Biziulk, M., J. Namiesnik, J. Czerwinski, D. Gorlo, B. Makuch, W. Janicki, Z. Polkowska, and L. Wolska. 1996. Occurrence and Determination of Organic Pollutants in Tap and Surface Waters of the Gdansk District. *J. Chromatogr. A*, 733(1-2): 171-183.

Blanchard, R. D., and J. K. Hardy. 1984. Use of a Permeation Sampler for Determination of Volatile Priority Pollutants. *Anal. Chem.*, 56(9): 1621-1624.

Brodbelt, J. S., and R. G. Cooks. 1987. In Vivo Mass Spectrometric Determination of Organic Compounds in Blood With a Membrane Probe. *Anal. Chem.*, 59(-): 454-458.

Brush, R. C., and G. W. Rice. 1994. Trihalomethanes Produced in Humic Acid Reactions—a GC-MS Experiment for Instrumental Analysis. *J. Chem. Educ.*, 71(12): A293-296.

Budde, W. L. 2001. *Analytical Mass Spectrometry.* New York: Oxford University Press.

Cai, H., S. D. Stearns, and W. E. Wentworth. 1998. Pulsed Discharge Electron Capture Detector Operating in the Constant-Current Mode by Means of Feedback dc Bias Voltage. *Anal. Chem.*, 70(18): 3770-3776.

Cancho, B., F. Ventura, and M. T. Galceran. 1999. Behavior of Halogenated Disinfection By-Products in the Water Treatment Plant of Barcelona, Spain. *B. Environ. Contam. Tox.*, 63(5): 610-617.

Carrero, H., and J. F. Rusling. 1999. Analysis of Haloacetic Acid Mixtures by HPLC Using Electrochemical Detector Coated With Surfactant-Nafion Film. *Talanta*, 48(3): 711-718.

Chang, C. C., and G. R. Her. 2000. On-line Monitoring Trihalomethanes in Chlorinated Water by Membrane Introduction-Fast Gas Chromatography Mass-Spectrometry. *J. Chromatogr. A*, 893(1): 169-175.

Chen, M.-L., I.-F. Mao, and C.-C. Hsu. 1997. Determination of Trihalomethanes in Drinking Water by Solid-Phase Microextraction and Gas Chromatography With Electron Capture Detection. *Toxicol. Environ. Chem.*, 60(1-4): 39-45.

Chen, T.-C., and G.-R. Her. 2001. On-line Monitoring of Trihalomethanes in Drinking Water Using Continuous-Flow Purge and Cryofocusing Gas Chromatography-Mass Spectrometry. *J. Chromatogr. A*, 927(1-2): 229-235.

Cho, D.-H., S.-H. Kong, and S.-G. Oh. 2003. Analysis of Trihalomethanes in Drinking Water Using Headspace-SPME Technique With Gas Chromatography. *Wat. Res.*, 37(2): 402-408.

Clemens, M., and H. F. Schoeler. 1992. Determination of Halogenated Acetic Acids and 2,2-dichloropropionic Acid in Water Samples. *Fresen J. Anal. Chem.*, 344(1-2): 47-49.

Dean, J. A. 1985. *Lange's Handbook of Chemistry.* 13th ed. New York: McGraw Hill.

Donnell, G. E., A. Juska, R. Geyer, M. Faiz, and S. Stalder. 1995. Analysis of Trichloroacetic Acid in the Urine of Workers Occupationally Exposed to Trichloroethylene by Capillary Gas Chromatography. *J. Chromatogr. A*, 709(-): 313-317.

Drechsel, D., K. Dettmer, W. Engewald, T. Bittner, and J. Efer. 2001. GC Analysis of Trichloroacetic Acid in Water Samples by Large-Volume Injection and Thermal Decarboxylation in a Programmed-Temperature Vaporizer. *Chromatographia*, 54: 151-154.

Duty, C. N. 2000. Application of a Supported Capillary Membrane Sampler for the Analysis of Organic Compounds in Water. Master's thesis. Arkansas State University, State University, Arkansas.

Ells, B., D. A. Barnett, K. Froese, R. W. Purves, S. Hrudey, and R. Guevremont. 1999. Detection of Chlorinated and Brominated Byproducts of Drinking Water Disinfection Using Electrospray Ionization-High-Field Asymmetric Waveform Ion Mobility Spectrometry-Mass Spectrometry. *Anal. Chem.*, 71(20): 4747-4752.

Ells, B., D. A. Barnett, R. W. Purves, and R. Guevremont. 2000. Detection of Nine Chlorinated and Brominated Haloacetic Acids at Part-per-Trillion Levels Using ESI-FAIMS-MS. *Anal. Chem.*, 72(19): 4555-4559.

Gabryelski, W., F. Wu, and K. L. Froese. 2003. Comparison of High-Field Asymmetric Waveform Ion Mobility Spectrometry With GC Methods in Analysis of Haloacetic Acids in Drinking Water. *Anal. Chem.*, 75(10): 2478-2486.

Gelover, S., E. R. Bandala, T. Leal-Ascencio, S. Perez, and E. Martinez. 2000. GC-MS Determination of Volatile Organic Compounds in Drinking Water Supplies in Mexico. *Environ. Toxicol.*, 15(2): 131-139.

Glaser, J. A., and D. L. Forest. 1981. Trace Analysis of Wastewaters. *Environ. Sci. Technol.*, 15: 1426-1435.

Global FIA Inc. 1999. *Supported Capillary Membrane Sampler Operation Manual.* Fox Harbor, Wash.

Gremaud, G., and W. E. Wentworth. 1996. Windowless Pulsed Discharge Photoionization Detector: Application to Quantitative Analysis of Volatile Organic Compounds. *J. Chromatogr. A*, 724(-): 235-250.

Guidotti, M., and G. Ravajoli. 1999. Analysis of Organic Micro Pollutants in Drinking Water Using SPME and GC-MS. *Annali di Chimica*, 89(11-12): 919-924.

Hashimoto, S., and A. Otsuki. 1998. Simultaneous Determination of Haloacetic Acids in Environmental Waters Using Electrospray Ionization Liquid Chromatography Mass Spectrometry. *J. High Resol. Chromatogr.,* 21(1): 55-58.

Hozalski, R.M., and L. Zhang. 2002. *Development of a Safe and Rapid Method for Analysis of Haloacetic Acids in Drinking Water.* [Draft.] Denver, Colo.: AwwaRF.

Humbert, B. E., and J. G. Fernandez. 1976. Simultaneous Determination of Trichloroacetic Acid and Trichloroethanol by Gas Chromatography. *Int. Arch. Occup. Environ. Health*, 36(4): 235-241.

Husain, S., R. Narshima, S. N. Alvi, and R. N. Rao. 1992. Monitoring the Effluents of the Trichloroacetic Acid Process by High-Performance Liquid Chromatography. *J. Chromatogr. A*, 600(2): 316-319.

Kim, D.-H., J.-O. Choi, M. Kim, and D. W. Lee. 2001. Determination of Haloacetic Acids in Tap Water by Capillary Electrophoresis With Direct UV Detection. *J. Liq. Chromatogr. R.T.*, 24(1): 47-55.

Kotiaho, T., T. K. Choudhury, R. G. Cooks, and F. R. Lauritsen. 1991. Membrane Introduction Mass Spectrometry. *Anal. Chem.*, 63(18): 875-883.

Kristiansen, N. K., K. T. Aune, M. Frøshaug, G. Becher, and E. Lundanes. 1996. Determination of Halogenated Acetic Acids in Chlorinated Sea Water and Drinking Water Produced Offshore. *Wat. Res.*, 30(9): 2155-2159.

Kuivinen, J., and H. Johnsson. 1999. Determination of Trihalomethanes and Some Chlorinated Solvents In Drinking Water By Headspace Technique With Capillary Column Gas Chromatography. *Wat. Res.*, 33(5): 1201-1208.

Kuklenyik, Z., D. L. Ashley, and A. M. Calafat. 2002. Quantitative Detection of Trichloroacetic acid in Human Urine Using Isotope Dilution High-Performance Liquid Chromatography-Electrospray Ionization Tandem Mass Spectrometry. *Anal. Chem.*, 74(9): 2058-2063.

Lauritsen, F. R., L. E. Dejarme, R. G. Cooks, and T. K. Choudhury. 1992. Micro Porous Membrane Introduction Mass Spectrometry With Solvent Chemical Ionization and Glow Discharge for the Direct Detection of Volatile Organic Compounds in Aqueous Solutions. *Anal. Chim. Acta*, 266(-): 1-12.

Liao, Z. 2001. Development of a Supported Capillary Membrane Sampler-Gas Chromatograph for the Analysis of Volatile Organic Compounds in Water. Master's thesis. Arkansas State University, State University, Arkansas.

Liu, Y., and S. Mou. 2003. Determination of Trace Levels of Haloacetic Acids and Perchlorate in Drinking Water by Ion Chromatography with Direct Injection. *J. Chromatogr. A*, 997 (1-2): 225-235.

Liu, Y., S. Mu, A. Lin, C. Jianhua, B. Du, and J. Cui. 2003. Determination of Nine Haloacetic Acids and Six Anions in Drinking Water by Ion Chromatography with Large Volume Direct Injection. *Chinese J. Chromatogr.*, 21(2): 181-183.

Loos, R., and D. Barcelo. 2001. Determination of Haloacetic Acids in Aqueous Environments by Solid-phase Extraction Followed by Ion-Pair Liquid Chromatography-Electrospray Ionization Mass Spectrometric Detection. *J. Chromatogr. A*, 938(1-2): 45-55.

Lord, T. 1999. Determination of Trihalomethanes (THMs), Trihaloacetic Acids (THAAs) and Other Disinfection By-products (DBPs) in Drinking Water. In *Proc. of the AWWA Water Quality Technology Conference.* Denver, Colo.: AWWA.

Lopez-Avila. 2000. *Development of an Experimental Procedure for Haloacetic Acids Using Ion Chromatography.* Denver, Colo.: AwwaRF.

Lopez-Avila, V., J. Benedicto, H. Prest, and S. Bauer. 1999. Automated MIMS for Direct Analysis of Organic Compounds in Water. *Am. Lab.*, 31(12): 32-37.

Magnuson, M. L., and C. A. Kelty. 2000. Microextraction of Nine Haloacetic Acids in Drinking Water at Microgram per Liter Levels With Electrospray-Mass Spectrometry of Stable Association Complexes. *Anal. Chem.*, 72(10): 2308-2312.

Martinez, D., F. Borrull, and M. Calull. 1998. Comparative Study of a Solid-Phase Extraction System Coupled to Capillary Electrophoresis in the Determination of Haloacetic Compounds in Tap Water. *J. Chromatogr. A*, 827(1): 105-112.

Martinez, D., F. Borrull, and M. Calull. 1999a. Evaluation of Different Electrolyte Systems and On-line Preconcentrations for the Analysis of Haloacetic Acids by Capillary Zone Electrophoresis. *J. Chromatogr. A*, 835(1+2): 187-196.

Martinez, D., F. Borrull, and M. Calull. 1999b. Strategies for Determining Environmental Pollutants at Trace Levels in Aqueous Samples by Capillary Electrophoresis. *Trends Anal. Chem.*, 18(4): 282-291.

Martinez, D., F. Borrull, M. Calull, J. Ruana, and A. Colom. 1998a. Application of Solid-phase Extraction Membrane Disks in the Determination of Haloacetic Acids in Water Using Gas Chromatography-Mass Chromatography. *Chromatographia*, 48(11/12): 811-816.

Martinez, D., J. Farre, F. Borrull, M. Calull, J. Ruana, and A. Colom. 1998b. Capillary Zone Electrophoresis With Indirect UV Detection of Haloacetic Acids in Water. *J. Chromatogr. A*, 808(-): 229-236.

McKenzie, K., W. Studabaker, and K. Carter. 1997. A Rapid, Cost-Effective Method for Detection of Trihalomethanes in Drinking Water. In *Proc. of the AWWA Water Quality Technology Conference*, 21/A-30/A(-). Denver, Colo.: AWWA.

Mendonca, S., and W. E. Wentworth. 1996. Relative Responses of Various Classes of Compounds Using a Pulsed Discharge Helium Ionization Detector: Experimental Determination and Theoretical Calculations. *J. Chromatogr. A*, 749(-): 131-148.

Nair, L. M., R. Saari-Nordhaus, and J. M. Anderson. 1994. Determination of Haloacetic Acids by Ion Chromatography. *J. Chromatogr. A*, 671(1-2): 309-313.

Nicholson, B. C., B. P. Maguire, and D. B. Bursill. 1984. Henry's Law Constants for the Trihalomethanes: Effect of Water Composition and Temperature. *Environ. Sci. Technol.*, 18(7): 518-521.

Nikolaou, A. D., S. K. Golfinopoulos, M. N. Kostopoulou, and T. D. Lekkas. 2002a. Determination of Haloacetic Acids in Water by Acidic Methanol Esterification-GC-ECD Method. *Wat. Res.*, 36(4): 1089-1094.

Nikolaou, A. D., T. D. Lekkas, S. K. Golfinopoulos, and M. N. Kostopoulou. 2002b. Application of Different Analytical Methods for Determination of Volatile Chlorination By-products in Drinking Water. *Talanta*, 56(4): 717-726.

Nilsson, T., F. Pelusio, L. Montanarella, B. Larsen, S. Facchetti, and J. Madsen. 1995. An Evaluation of Solid Phase Microextraction for Analysis of Volatile Organic Compounds in Drinking Water. *J. High Resol. Chromatogr.*, 18(10): 617-624

Ogata, M., Y. Takatsuka, and K. Tomokuni. 1970. A Simple Method for the Quantitative Analysis of Urinary Trichloroethanol and Trichloroacetic Acid as an Index of Trichloroethylene Exposure. *Brit. J. Industr., Med.*, 27(4): 378-381.

Ozawa, H. 1993. Gas Chromatographic-Mass Spectrometric Determination of Halogenated Acetic Acids in Water After Direct Derivatization. *J. Chromatogr. A*, 644(2): 375-382.

Pal, A. 1996. Spectrophotometric Determination via Photochemical Reaction of Trichloroacetic Acid and Its Application for Water Analysis. *Chem ANAL-WARSAW*, 41(1): 65-69.

Pillai, A., R. Rastogi, and V. K. Gupta. 1999. A New Spectrometric Method for Determination of Trichloroacetic Acid in Air and Environmental Samples. *J. Indian Chem. Soc.*, 76(9): 467-468.

Pratt, K. F., and J. Pawliszyn. 1992a. Gas Extraction Kinetics of Volatile Organic Species From Water With a Hollow Fiber Membrane. *Anal. Chem.*, 64(18): 2101-2106.

Pratt, K. F., and J. Pawliszyn. 1992b. Water Monitoring System Based on Gas Extraction With a Single Hollow Fiber Membrane and Gas Chromatographic Cryotrapping. *Anal. Chem.*, 64(18): 2107-2110.

Rathore, H. S., S. K. Saxena, and T. Begum. 1992. A Selective Fluorescence Spot Test for the Detection of TCA in Soil and Water. *J. Indian Chem. Soc.*, 69(11): 798-799.

Reckhow, D. A., and P. D. Pierce. 1992. *A Simple Spectrometric Method for the Determination of THMs in Drinking Water.* Denver, Colo.: AwwaRF and AWWA.

Reemtsma, T. 2001. The Use of Liquid Chromatography-Atmospheric Pressure Ionization-Mass Spectrometry in Water Analysis—Part 1: Achievements. *Trends Anal. Chem.*, 20(9): 500-517.

Richardson, S. D. 2001a. Mass Spectrometry in Environmental Sciences. *Chemical Reviews, Washington, D.C.*, 101(2): 211-254.

Richardson, S. D. 2001b. Water Analysis. *Anal. Chem.*, 73(12): 2719-2734.

Richardson, S. D. 2002. The Role of GC-MS and LC-MS in the Discovery of Drinking Water Disinfection By-Products. *J. Environ. Monit.*, 4(1): 1-9.

Roehl, R., R. Slingsby, N. Avdalovic, and P. E. Jackson. 2002. Application of Ion Chromatography With Electrospray Mass Spectrometric Detection to the Determination of Environmental Contaminants in Water. *J. Chromatogr. A*, 956(1-2): 245-254.

Sarrion, M. N., F. J. Santos, and M. T. Galceran. 1999. Solid-phase Microextraction Coupled With Gas Chromatography-Ion Trap Mass Spectrometry for the Analysis of Haloacetic Acids in Water. *J. Chromatogr. A*, 859(2): 159-171.

Sarrion, M. N., F. J. Santos, and M. T. Galceran. 2000. In Situ Derivatization/Solid-phase Microextraction for the Determination of Haloacetic Acids in Water. *Anal. Chem.*, 72(20): 4865-4873.

Sarzanini, C., M. C. Bryzzoniti, and E. Mentasti. 1999. Preconcentration and Separation of Haloacetic Acids by Ion Chromatography. *J. Chromatogr. A*, 850(1+2): 197-211.

Satinsky, D., H. Sklenarova, J. Huclova, and R. Karlicek. 2003. On-line Coupling of Sequential Injection Extraction With Restricted-Access Materials for Sample Clean-up and Analysis of Drugs in Biological Matrix. *Analyst*, 128(4): 351-356.

Scott, B. F., and M. Alaee. 1998. Determination of Haloacetic Acids From Aqueous Samples Collected From the Canadian Environment Using an In Situ Derivatization Technique. *Water Qual. Res. J. Canada*, 33(2): 279-293.

Scott, B. F., D. Mactavish, C. Spencer, W. M. J. Strachan, and D. C. G. Muir. 2000. Haloacetic Acids in Canadian Lake Waters and Precipitation. *Environ. Sci. Technol.*, 34(20): 4266-4272.

Sigma Aldrich Inc. 2003. SPME Literature [Online]. Available: <http://www.sigmaaldrich.com/Brands/Supelco_Home/Spotlights/SPME_central.html>. [cited July 29, 2003]

Slivon, L. E., J. S. He., W. L. Budde, and M. R. Bauer. 1991. Helium Purged Hollow Fiber Membrane Mass Spectrometer Interface for Continuous Measurement of Organic Compounds in Water. *Anal. Chem.*, 63(-): 1335-1340.

Soga, T., and M. Imaizumi. 2001. Capillary Electrophoresis Method for the Analysis of Inorganic Anions, Organic Acids, Amino Acids, Nucleotides, Carbohydrates and Other Anionic Compounds. *Electrophoresis*, 22(16): 3418-3425.

Stack, M. A., G. Fitzgerald, S. O'Connell, and K. J. James. 2000. Measurement of Trihalomethanes in Potable and Recreational Waters Using Solid-Phase Micro Extraction With Gas Chromatography-Mass Spectrometry. *Chemosphere*, 41(11): 1821-1826.

Staples, E. J. 2000. The zNose, A New Electronic Nose Using Acoustic Technology. *Acoustical Society of America*, (2aEA4).

St. Germain, F., O. Mamer, J. Brunet, B. Vachon, R. Tardif, T. Abribat, C. D. Roisers, and J. Montgomery. 1995. Volatile Organic Compound Analysis by an Inertial Spray Extraction Interface Coupled to an Ion Trap Mass Spectrometer. *Anal. Chem.*, 67(24): 4536-4541.

Stuetz, R. 2001. Using Sensor Arrays for On-line Monitoring of Water and Wastewater Quality. *Am. Lab.*, 33(2): 10,12,14.

Takahashi, Y., S. Onodera, M. Morita, and Y. Terao. 2003. A Problem in the Determination of Trihalomethanes by Headspace-Gas Chromatography/Mass Spectrometry. *J. Health. Sci.*, 49(1): 1-7.

Takino, M., S. Daishima, and K. Yamaguchi. 2000. Determination of Haloacetic Acids in Water by Liquid Chromatography-Electrospray Ionization-Mass Spectrometry Using Volatile Ion-Pairing Reagents. *Analyst*, 125(6): 1097-1102.

Tanaka, S., and M. Ikeda. 1968. A Method for Determination of Trichloroethanol and Trichloroacetic Acid in Urine. *Brit. J. Industr., Med.*, 25(3): 214-219.

Urbansky, E. T. 2000a. Disinfection By-Products in Drinking Water. *Anal. Chem.*, 72(13): 439A-440A.

Urbansky, E. T. 2000b. Techniques and Methods for the Determination of Haloacetic Acids in Potable Water. *J. Environ. Monit.*, 2(5): 285-291.

USEPA (U.S. Environmental Protection Agency). 1990a. Method 551: *Determination of Chlorinated Disinfection By-products and Chlorinated Solvents in Drinking Water by Liquid Liquid Extraction and Gas Chromatography with Electron Capture Detector*, Cincinnati, Ohio: Environmental Monitoring and System Laboratory.

USEPA (U.S. Environmental Protection Agency). 1990b. Method 552: *Determination of Haloacetic Acids in Drinking Water by Liquid-Liquid Extraction, Derivatization and Gas Chromatography With Electron Capture Detector*, Cincinnati, Ohio: Environmental Monitoring and System Laboratory.

USEPA (U.S. Environmental Protection Agency). 1992. Method 552.1: *Determination of Haloacetic Acids and Dalapon in Drinking Water by Ion Exchange Liquid Solid Extraction, Derivatization and Gas Chromatography With Electron Capture Detection*, Cincinnati, Ohio: Environmental Monitoring and System Laboratory.

USEPA (U.S. Environmental Protection Agency). 1995a. Method 502.2: *Volatile Organic Compounds in Water by Purge and Trap Capillary Gas Chromatography with a Photoionization and Electrolytic Conductivity Detector in Series*, Cincinnati, Ohio: Environmental Monitoring and System Laboratory.

USEPA (U.S. Environmental Protection Agency). 1995b. Method 524.2: *Measurement of Purgeable Organic Compounds in Water by Capillary Column Gas Chromatography / Mass Spectrometry*, Cincinnati, Ohio: Environmental Monitoring and System Laboratory.

USEPA (U.S. Environmental Protection Agency). 1995c. Method 551.1: *Determination of Chlorinated Disinfection Byproducts, Chlorinated Solvents, and Halogenated Pesticides / Herbicides in Drinking Water by Liquid Liquid Extraction and Gas Chromatography with Electron Capture Detector*, Cincinnati, Ohio: Environmental Monitoring and System Laboratory.

USEPA (U.S. Environmental Protection Agency). 1995d. Method 552.2: *Dalapon in Drinking Water by Ion Exchange Liquid Solid Extraction, Derivatization and Gas Chromatography With Electron Capture Detector*, Cincinnati, Ohio: Environmental Monitoring and System Laboratory.

USEPA (U.S. Environmental Protection Agency). 1996. *DBP/ICR Analytical Methods Manual.* USEPA, Office of Water. EPA 814-B-96-002.

USEPA (U.S. Environmental Protection Agency). 1998. National Primary Drinking Water Regulations: Disinfectants and Disinfection By-Products Rule: Final Rule. *Fed. Reg.*, 63, 241: 68390.

USEPA (U.S. Environmental Protection Agency). 2003a. Appendix B, Part 136—Definition and Procedure for the Determination of the Method Detection Limit—Revision 1.11. *Fed. Reg.*

USEPA (U.S. Environmental Protection Agency). 2003b. Method 552.3: *Determination of Haloacetic Acids and Dalapon in Drinking Water by Liquid-Liquid Microextraction, Derivatization and Gas Chromatography With Electron Capture Detector*, Cincinnati, Ohio: Environmental Monitoring and System Laboratory.

USEPA (U.S. Environmental Protection Agency). 2003c. Stage 2 Disinfectant-Disinfection By-products Rule-Significant Excursion Guidance Manual [Draft] [Online]. Available: <http://www.epa.gove/safewater/stage2/index.html>. [cited July 30,2003]

Vasnin, S. V., and W. E. Wentworth. 1992. Pulsed Discharge Emission Detector-Application to Analytical Spectroscopy of Permanent Gases. *Chromatographia*, 34(5-8): 219-225.

Vesterberg, O., J. Gorczak, and M. Krasts. 1975. Methods for Measuring Trichloroethanol and Trichloroacetic acid in Blood and Urine After Exposure to Trichloroethylene. *Scand. J. Work Environ. & Health*, 1(4): 243-248.

Vichot, R., and K. G. Furton. 1994. Variables Influencing the Direct Determination of Haloacetic Acids in Water by Reversed-Phase Ion-Pair Chromatography With Indirect UV Detection. *J. Liq. Chromatogr.*, 17(20): 4405-4429.

Vogt, F., M. Karlowatz, M. Jakusch, and B. Mizaikoff. 2003. The Automated Sample Preparation System Mix Master for Investigation of Volatile Organic Compounds With Mid-Infrared Evanescent Wave Spectroscopy. *Analyst*, 128(4): 397-403.

Wang, H., M. S. Simmons, and R. A. Deininger. 1995. Application of a Dynamic Headspace Analysis for Trihalomethanes in Flowing Water. *J. Chromatogr. Sci.*, 33(-): 109-115.

Wang, K., and S. Tan. 2000. Solid Phase Microextraction and GC-MS Analysis of Trihalomethanes in High-purity Water. *Ultra Pure Water*, 17(6): 55-59.

Weinberg, H. S., S. W. Krasner, S. Richardson, and R. Sangaiah. 1999. Current Research into the Occurrence of New Disinfection By-Products in Drinking Water. In *Book of Abstracts*, 217th ACS National Meeting, March 21-25, Anaheim, Calif.

Wentworth, W. E., and H. Cai. 1994. Pulsed Discharge Helium Ionization Detector: Universal Detector for Inorganic and Organic Compounds at the Low Picogram Level. *J. Chromatogr.*, 688(1+2): 135-152.

Wentworth, W. E., and E. D. D'sa. 1992. Environmental Applications of the Pulsed Discharge Electron Capture Detector. *J. Chromatogr. Sci.*, 30(12): 478-485.

Wentworth, W. E., and T. Limero. 1989. Non-Radioactive Electron Capture Detector for Gas Chromatography. *J. Chromatogr.*, 468(-): 215-224.

Wentworth, W. E., Y. Wang, W. Odegard, E. C. M. Chen, and S. D. Stearns. 1996. Pulsed-Discharge Electron-Capture Detector: Kinetic Model, Response Factors, and Temperature Dependence. *J. Chromatogr. Sci.*, 34(8): 368-375.

Wentworth, W. E., and S. Watanesk. 1996. Experimental and Theoretical Relative Response Factors for Pulsed Discharge Krypton Photoionization Detector. *J. Chromatogr.*, 749(1+2): 149-155.

Westover, L. B., and J. C. Tou. 1974. Novel Mass Spectrometric Sampling Device Hollow Fiber Probe. *Anal. Chem.*, 46(4): 568-571.

Wolcott, W., and Olson, D. (In review.) Application for a US patent. Sequential Injection Liquid-Liquid Extraction.

Wong, P. S. H., and R. G. Cooks. 1995. Trace Level Analysis of Volatile and Semivolatile Organic Compounds in Water Using a Membrane/Jet Separator Interfaced to an Ion-trap Mass Spectrometer. *Analytica Chemica Acta*, 31(-): 387-398.

Wu, F., W. Gabryelski, and K. Froese. 2002. Improved Gas Chromatography Methods for Micro-Volume Analysis of Haloacetic Acids in Water and Biological Matrices. *Analyst*, 127(10): 1318-1323.

Xie, Y. 2001. Analyzing Haloacetic Acids Using Gas Chromatography/Mass Spectrometry. *Wat. Res.*, 35(6): 1599-1602.

Xie, Y., D. A. Reckhow, and D. C. Springborg. 1998. Analyzing HAAs and Ketoacids Without Diazomethane. *Jour. AWWA*, 90(4): 131-138.

Xie, Y., H. J. Zhou, and J. P. Romano. 2000. Development of a Capillary Electrophoresis Method for Haloacetic Acids. *ACS*, 761(-): 356-365.

Yinon, J. 2003. Detection of Explosives by Electronic Noses. *Anal. Chem.*, 75(5): 98A-105A.

Zhang, L., X. Qian, and G. Yang. 2001. Determination of Trihalomethanes in Water by Purge-and-Trap Gas Chromatography With Mass Spectrometry and Electron Capture Detection. In *Preprints of Extended Abstracts.* Presented at the ACS National Meeting, American Chemistry Society, Division of Environmental Chemistry, 41(1): 18-22.

Zygmunt, B. 1996. Determination of Trihalomethanes in Aqueous Samples by Means of a Purge-and-Trap System With On-Sorbent Focusing Coupled to Gas Chromatography With Electron-Capture Detection. *J. Chromatogr. A*, 725 (1): 157-163.

ABBREVIATIONS

ADS	alkyl-diol-silica
AEE	acidic ethanol esterification
Ag/AgCl	silver/silver chloride
AME	acidic methanol esterification
Amu	atomic mass units
ANOVA	analysis of variance
Ar/CH_4	argon/methane
AWWA	American Water Works Association
AwwaRF	Awwa Research Foundation
°C	degrees Celsius (centigrade)
$Ca(OCl)_2$	calcium hypochlorite
CAR	carboxen
CE	capillary electrophoresis
$CHBr_3$	bromoform
$CHBrCl_2$	bromodichloromethane
$CHBr_2Cl$	dibromochloromethane
$CHCl_3$	chloroform
CID	collision induced dissociation
Cl^-	chloride ion
Cl_2	chlorine
cm	centimeter
cm^2	square centimeter
°C/min	degrees Celsius per minute
cm/s	centimeters per second
CO_2	carbon dioxide
CTAB	cetyltrimethylammonium bromide
CTAC	cetyltrimethylammonium chloride
CTAH	cetyltrimethylammonium hydrohide
CZE	capillary zone electrophoresis
DAI	direct aqueous injection
DBA	dibutylamine
DBAA	dibromoacetic acid
1,3-DBB	1,3-dibromobutane
2,3-DBPA	2,3-dibromopropionic acid
DBPs	disinfection by-products
DCAA	dichloroacetic acid
2,3-DCPA	2,3-dichloropropionic acid
DI	deionization
DIP	direct insertion probe
DMS	dimethylsiloxane

DNT	dinitrotoluene
DVB	divinylbenzene
ECD	electron capture detector
EI	electron impact ionization
e.l.	effective length
EMV	electron multiplier voltage
EOF	electro-osmotic flow
ESI	electrospray ionization
eV	electronvolt
FAIMS	high-field asymmetric waveform ion mobility mass spectrometry
FIA	flow injection analysis
FID	flame ionization detector
f.t.	film thickness
ft	foot
FTIR	Fourier transform infrared spectroscopy
GC	gas chromatography
H_2	hydrogen
HAAs	haloacetic acids
Hall	electrolytic conductivity detector
He	helium
HP	Hewlett-Packard
HPLC	high performance liquid chromatography
hr	hour
HS	headspace sampling
IC	ion chromatography
i.d.	internal diameter
in.	inch
IT	ion trap
J.	Journal
Jour.	Journal
K	distribution constant
KOH	potassium hydroxide
kPa	kilopascal
kV	kilovolt
L	liter
LC	liquid chromatography
LLE	liquid-liquid extraction

LOQ	limit of quantification
LVI	large volume injection
λ	wavelength
M	moles per liter
MBAA	monobromoacetic acid
Mbar	millibar
MCAA	monochloroacetic acid
MCL	maximum contaminant level
MDL	method detection limit
MDQ	minimum detectable quantity
MEKC	micellar electrokinetic capillary chromatography
MeOH	methanol
MHz	megahertz
MIMS	membrane introduction spectrometry
min.	minute
mL	milliliter
mL/min	milliliters per minute
Mm	millimeter
mM/L	millimoles per liter
MS	mass spectrometry
MSD	mass selective detector
Msec	millisecond
MTBE	methyl tert-butyl ether
m/z	mass to charge ratio
MΩ	mega ohm
µA	microampere
µg/L	micrograms per liter
µm	micrometer
N_2	nitrogen
NaCl	sodium chloride
NaOCl	sodium hypochlorite
NaOH	sodium hydroxide
NDC	naphthaline dicarboxylic acid dipotassium salt
ng/mL	nanograms per milliliter
Nm	nanometer
NOM	natural organic matter
o.d.	outer diameter
PA	polyacrylate
PAC	Project Advisory Committee
PDC	2,6-pyridinedicarboxylic acid
PDECD	pulsed discharge electron capture detector
PDHID	pulsed discharge helium ionization detector

PDMS	polydimethylsiloxane
PDPID	pulsed discharge photoionization detector
PGA	programmable gate array
pH	negative logarithm of the effective hydrogen-ion concentration
PID	photoionization detector
PPy/SBA	polypyrrole/sulfobenzoic acid
Psi	pounds per square inch
psi/min	pounds per square inch per minute
P&T	purge and trap
PTFE	polytetrafluoroethylene
PTV	programmed-temperature vaporization
QAPP	Quality Assurance Project Plan
r	Pearson correlation coefficient
r^2	Pearson determination coefficient
RAM	restricted access material
RF	radio frequency
RO	reverse osmosis
% RSD	percent relative standard deviation
SAW	surface acoustic wave
SAX	strong anionic exchanger
SCMS	supported capillary membrane sampling
s/cycle	scans per cycle
SIA	sequential injection analysis
SILEX	sequential injection liquid-liquid extraction
SIM	selected ion monitoring
S/N	signal to noise ratio
SPE	solid phase extraction
SPME	solid phase microextraction
SSI	Scientific System Inc.
T	absolute temperature
TBAA	tribromoacetic acid
TBA-HSO_4	tetrabutylammonium bromide
TCAA	trichloroacetic acid
THEED	transmission high energy electron diffraction
THM	trihalomethane
TTAB	tetradecyltrimethylammonium bromide
TTHM	total trihalomethane
USEPA	U.S. Environmental Protection Agency
UV	ultraviolet

V	volt
VIS	visible
VOCs	volatile organic compounds
v/v	volume per volume
Xe	xenon
XRD	x-ray diffractometry

IWA Publishing's authorised EU representative for General Product Safety Regulations is Diane D'Arras, 15 rue Duret, 75116 Paris, France, e-mail: safety@iwap.co.uk.

Printed and bound by CPI Group (UK) Ltd, Croydon, CR0 4YY
17/09/2026
02218793-0017